Comprehensive Analytical Chemistry
Volume 72

Rapid Immunotests for Clinical, Food and Environmental Applications

Advisory Board

Comprehensive Analytical Chemistry
Volume 72

Rapid Immunotests for Clinical, Food and Environmental Applications

Irina Yu. Goryacheva
Saratov State University,
Saratov, Russia

ELSEVIER

AMSTERDAM • BOSTON • HEIDELBERG • LONDON
NEW YORK • OXFORD • PARIS • SAN DIEGO
SAN FRANCISCO • SINGAPORE • SYDNEY • TOKYO

Elsevier
Radarweg 29, PO Box 211, 1000 AE Amsterdam, Netherlands
The Boulevard, Langford Lane, Kidlington, Oxford OX5 1GB, UK
50 Hampshire Street, 5th Floor, Cambridge, MA 02139, USA

ISBN: 978-0-444-63574-7
ISSN: 0166-526X

Publisher: Zoe Kruze
Acquisition Editor: Poppy Garraway
Editorial Project Manager: Shellie Bryant
Production Project Manager: Radhakrishnan Lakshmanan
Cover Designer: Alan Studholme

Typeset by TNQ Books and Journals

Contents

Series Editor's Preface

Screening techniques is an emerging issue in modern analytical chemistry due to the need to obtain direct and fast analytical information. The importance of such an area has been already recognized by us by publishing in this series volume 36 on Chemical Test Methods of Analysis from Yu.A. Zolotov, V.M. Ivanov and V.G. Amelin and volume 39 on Integrated Analytical Systems edited by my friend and colleague Salvador Alegret. The publication of these volumes goes back to 2003, so now more than 10 years after we publish a book on Rapid Immunotests. It does not mean at all that I was not interested in that topic. Indeed I always felt it was of great importance to the analytical chemistry community, but unfortunately it took me quite a while to find the appropriate editor to edit a book.

The present book by Irina Yu. Gorvacheva complements very well previous books, and it is an additional step in the area of rapid analysis. It offers a comprehensive study on the fundamentals, formats, materials, labels and multifunctionality of immunotests. The book contains seven chapters summarizing the main methodologies and achievements in the field of rapid immunotests in analytical chemistry up to date.

Overall this book covers most of the immunotest-based technological advances applied to analytical chemistry that are the focus of today's attention. I am convinced that this book will be of great help to newcomers who want to learn more about rapid test. But this book will be a guidance also for interdisciplinary experts interested in combining physical chemistry, microelectronics, biological sciences, analytical chemistry and automation. Finally, I would like to thank Irina Yu. Gorvacheva and also, those that have contributed in compiling such a unique book on rapid immunotests.

Prof. D. Barceló
Barcelona, Spain, 21 March 2016

Preface

Rapid methods occupy a fitting place in modern life, research and industry. Progress in this area is related with achievements in such traditional areas as biology, chemistry and physics as well as nanotechnology and material science. The scope of this volume is to provide an overview of the past, the present and look to the future of possibility of rapid non-laboratory detection and to list areas which could contribute to the development of rapid tests.

This book would be interesting for laboratory researchers and academics who work for development of immunoassay and nanosystems application, also for researchers who use immunoassay as a tool including clinical and veterinary chemists, biochemists and pathologists; also for students and researchers in nanomaterials, biochemistry and medicine. For students the sections about basic principles of tests would be useful.

An introduction chapter of the book pictures a rapid tests place among other analytical methods, the driving forces for development of rapid immunotests as well as nanotechnology investment into rapid test progress.

The second chapter is devoted to a historical retrospective of the development of principles which become to be a basis of modern rapid immunotests. Among the large variety of methods and tools, only those who brought principally new techniques for development of total conception of rapid tests are described. In this line, much attention has been paid to the radioimmunoassay, started a new era in the immunoassay but actually no longer used for research and routine monitoring. An attempt to trace the logic of the development was carried out and some curious (from modern point of view) things are presented. History of tests described more like popular scientific literature and could be interesting for the layman as well. Some methods, such as radioimmunoassay, that played an important role in the development of methodology, theoretical and practical aspects of assay, but for now are replaced by methods with better characteristics (safety, simplicity, sensitivity) are discussed in this chapter in detail.

The third chapter presents classification of rapid tests, description of principle and construction with maximum details, the current state of development of immunochemical tests for detection of low and high molecular weight analytes. For heterogeneous tests, classification is given on the base of solid support type: microtitre plates, thin-film membranes, bulk porous materials. The procedure and outcome of assay and label type are only playing an auxiliary role. For homogeneous assays, classification and order of the methods is given on the base of the label nature and properties because for this group of methods labels playing a role of active nanosensors to indicate an immunocomplex formation. The principle difference in label function for heterogeneous and homogeneous assays is discussed: the latter group label should be an active nanosensors only. For the first group both passive label and active nanosensors are employed; application of passive label is wider. Rapid enzyme-linked immunoassay tests, lateral flow and flow-through membrane and column immunoassays, immunoagglutination, fluorescence polarization immunoassays and Förster resonance energy transfer-based immunoassays are described with detailed schemes; possibilities to detect low and high molecular weight are discussed.

The forth chapter summarizes developments in optical labels for rapid immunotests. The latest achievements in materials science, in particular in the field of nanomaterials largely determine advancements in different areas of assays, including rapid immunotests. Principally new labels and improvement of characteristics and parameters of traditional ones, such as dyes and gold nanoparticles, took precedence in the list of labels of interest. Innovative and powerful novel labels have been developed in recent years. Basically studies related to labels development lies in bioimaging in cellular and molecular biology, biosensing and biotechnology. The combination of the antibodies' high specificity and ultrasensitive detection with advances of new labels is the most bright aspect in the field of rapid tests workflow. Nanoparticles coupled with biomolecules offer a very promising way to integrate biochemical recognition with inherent signal amplification. For each label the possibility of simultaneous detection of multiple analytes and reader application is discussed.

The fifth chapter summarizes developments in rapid immunotests multiplexing. Approaches for the simultaneous detection of multiple analytes are classified as the detection of multiple analytes in separate test-zones (with single and several labels) and within single test-zone. Microcarrier-based multiplex technologies and homogeneous multiplex assay are also described. Setting construction, restrictions, benefits and perspectives for simultaneous detection of multiple analytes in one probe with no or minimum sample preparation are discussed.

The sixth chapter summarizes hot-points in rapid tests development, such as an application of new substrate platforms with an example of photonic crystal fibres and multiplexing of labels with an example of multiloaded with quantum dots nanostructures. Photonic crystal fibres enable to trap light in a hollow void inside fibre by surrounding it with a 2D periodic photonic crystal. This allows turning of the waveguiding properties of the fibres. The second part of the chapter devoted to the construction of multiloaded with quantum dots labels. Nanoscale size of quantum dots offers the prospect of creating of multiloaded nanostructures to illustrate novel concepts and promising labels for analysis.

Such labels include carbon nanotubes decorated with quantum dots, layer-by-layer built microcapsules, loaded with quantum dots, dendrimer-based architectures with incorporated quantum dots and silica structures enveloped multiple quantum dots.

This is a book with a tiny bit of philosophy, a drop of history, little stories about rapid tests to describe the area from premier ABC-book level to the current situation and advanced technologies. With close attention to the basic principles and the specific issues, considering the breadth of the field that the rapid tests may offer, the coverage of this book is by no means complete, leaving room for challenge and research.

I would like to thank my friends and collaborators for their experience, help, comments, approval and disapproval, ideas and patience, flights of imagination and soft touchdowns. I would like to thank the series editor, Damia Barceló, for invitation to write this volume and the publishing team from Elsevier for their assistance. I would also like to thank my family, Alex and Olga, for their understanding, patience and support.

Irina Yu. Goryacheva
April 2016

Chapter 1

Introduction

There is a continuum in analytical technologies from methods, requiring specialists, infrastructure and a centralized approach, to less demanded technologies that can be used onsite and require little to no special knowledge and supporting infrastructure. Development of equipment-related techniques is the highest point of this trend, and for the last several decades humanity achieved a huge progress in this direction. A spectacular product of such development is a complex of mass spectrometry—based methods which is used to identify the amount and type of chemicals present in a sample by measuring the mass-to-charge ratio. These methods allow to implement things deemed fantastical just a few years ago, such as simultaneous detection of hundreds of substances at low concentration in nanolitre volume, discovery of new metabolites and confirmation of precursor structure, finding out the essence of unknown compounds, including the ones with molecular weight up to millions of Da, and establishing pathways and kinetics of reactions. Modern industry provided analytical techniques with fantastical possibilities to determine quality and quantity of substances and kinetics of processes using ion sources, orbitrap, triple quadrupoles and other fine tools.

On the other hand, for the last several million years nature has been dealing with the same problems. Which components are inside the body? Is concentration of them too high? Is their form and type correct and healthy? How to distinguish self and nonself? Nature developed complex systems of the living beings such as immune system and compliment system. Therefore, the second prominent trend in contemporary analytical science is to utilize interactions that have already been developed by nature for analytical purposes. This idea is especially attractive because such processes are very specific, very sensitive, run out at soft temperatures, in close to neutral aqueous medium in the presence of hundreds of various compounds. Thus, humanity needs only to find a way to make these processes useful for assay and to create settings for 'visualization' and 'understanding' of its outcomes. For this purpose it is necessary to modify these processes to make clearly distinguishable results for different concentration of target analyte (analytes). But the problem is how to make it visible to the naked eye or to the equipment. It can be solved by several means including by initiating a reaction resulting in distinctly detected phenomena, such as precipitation or colour changing, or by using special labels or equipments sensitive to the changes in medium properties in the

Comprehensive Analytical Chemistry, Vol. 72. http://dx.doi.org/10.1016/bs.coac.2016.03.006

presence of analyte. Application of biological components makes these processes very suitable for clinical assay and related areas, because of the initially physiological conditions of immunoreactions. Dry chemistry one-step tests, such as lateral flow immunoassays, have traditionally been placed at the lower end of the spectrum of analytical methods.

Today, immunoassays are applied everywhere from the clinical area to food and environment control, and now it is difficult to find a situation where immunoassay is not applicable. Rapid immunotests are appropriate for qualitative and semiquantitative analysis and also, for some extent, quantitative monitoring onsite or in resource-poor or nonlaboratory environments.

Driving forces for development of rapid immunotests include:

- Development of immunoreagents and methods of their synthesis.
- Development of new formats.
- Development of new settings to make existing laboratory methods applicable for out-of-laboratory conditions.
- Development of new equipment and readers, including those capable of detecting media properties (eg, emission decay) using compact, cheap, handheld instruments.
- Development of solid supports and platforms for immunoreactions, including platforms capable of signal modulation (eg, photonic crystal waveguides).
- Development of new labels.

The progress in developing cheap and fast detection tests is related to achievements in such high tech area as nanotechnology. The major merit of using nanotechnology is the possibility to control and tailor material properties in a very predictable manner to meet the needs of specific applications. A huge effort has been made in the field of rapid tests development to simplify the assay procedures while preserving the essential benefits such as sensitivity, robustness, broad applicability and suitability to automation. This has improved applicability of such tests in decentralized or out-of-lab environments. In order for this evolution in capability to occur, numerous improvements in such fields as labels, reagents, operating techniques, manufacturing equipment and process technology were made, and a whole new generation of facilitative technologies was introduced.

The impact of rapid tests progress for clinical, social, economic and other aspects of life can be suitably demonstrated by the progress made in pregnancy tests development since the 1970s. The first rapid pregnancy test called Wampole test became available to doctors and technicians only nearly 40 years ago. The test results were obtained via complex 2-h procedure. The first home pregnancy test called the e.p.t. (which originally stood for 'early pregnancy test', and later for more comforting 'error proof test') was introduced in 1977. This test also took 2 h and was more accurate when dealing with positive results. The chances of accuracy were 97% for positive results but only 80% for negative results.

In 1978, an issue of 'Mademoiselle' described the original e.p.t. as follows: 'For your $10, you get premeasured ingredients consisting of a vial of purified water, a test tube containing, among other things, sheep red blood cells as well as a medicine dropper and clear plastic support for the test tube, with an angled mirror at the bottom.' [Mademoiselle, April 1978, p. 86]. The advantages stated were, 'privacy and not having to wait several more weeks for the doctor's confirmation, which gives you a chance, if pregnant, to start taking care of yourself...'

Chapter 2

Rapid Tests Progress Through the Years

Chapter Outline

1. INTRODUCTION

The dream of magical rapid tests which allow you to immediately know the results without any tools and reagents accompanies humanity through the millennium. The attempts to clarify conviction of guilty wives with 'rice test', or determine gender of unborn child on the base of pregnant women urine properties are the first historical examples of such tests. Since then a lot was changed in methodology and validation, but it is interesting to mention the body fluids till now are the most usable matrices for rapid tests.

Despite best efforts through the ages, it was not until the mid-20th century that the majority of rapid methods gained real predictive value. The main key here was establishment of instrument for precise selection of an aim which is directly responsible for the target characteristic. The historical break-through was connected with introduction of antibody as recognizing tool and visualization of antibody–antigen interaction.

Prehistory of immunoanalytical methods is the development of immunology, started with the classical works of English physician Edward Jenner, demonstrated vaccination by injection of products of the sick organism as a mechanism to induce immunity to small pox and subsequently to other infectious diseases. In 1882 the Russian scientist Ilya Metchnikoff first proved that 'phagocytosis'; of microorganisms by leucocytes was the major host defence against infection [1]. In 1888 Emil von Behring first demonstrated that nonlethal doses of the toxin from diphtheria induced immunity to organism; soon after Emil von Behring and Shibasaburo Kitasato demonstrated that

Comprehensive Analytical Chemistry, Vol. 72. http://dx.doi.org/10.1016/bs.coac.2016.03.007

immunity to infection following injections of bacterial toxins was the result of the appearance of a 'factor' in the serum which neutralized the toxin [2]. This antitoxin activity could be transferred to normal animals by injecting serum from immune animals. It thus appeared that the body was able to respond to external infections by producing serum components that would neutralize their toxic effects. These serum components were called 'antibodies', and they escort the human and mammals for millions years protecting them against diseases. They are Y-shaped proteins produced by plasma cells that are used by the immune system to identify and neutralize foreign objects such as bacteria and viruses. The antibody binds the foreign aim, called an antigen by recognizing an antigen unique part, called antigen determinant or epitope. Epitope plays the role of 'key', compatible part of antibody (paratope) analogous to a lock. This key—lock bound is very precise and strong, while formed by big amount of weak intermolecular interactions. The role of an antibody in the blood is to tag a microbe or an infected cell for attack by other parts of the immune system, or neutralize it directly (eg, by blocking a part of a microbe that is essential for its invasion and survival).

The role of antibody as a tool for in vitro assay is also to capture and hold the target. Besides of traditional 'natural' antibody, different kinds of modified 'artificial' ones are developed for improving properties and simplification of procedure.

Another important element is the tool that would make antigen—antibody complex formation 'visible' for naked eyes and/or equipment/reader. The simplest effect to detect was the formation of the visible new phase, such as precipitate.

2. PRECIPITATION

The simplest immunoassay method is based on the detection of precipitate, which forms after the reagent antibody has incubated with the sample and reacted with their matching antigen to form insoluble aggregates [3]. The precipitation (precipitin test) is an example of a clinical test based on antigen—antibody reaction. Precipitation is a kind of serological reactions very similar to agglutination reaction though both are highly specific.

Michael Heidelberger in 1939 [3] developed the first time quantitative precipitation test and also purified antibody for the first time. In these experiments antibodies were able to precipitate antigens through multivalent binding, in which two Fab fragments in a single antibody can simultaneously bind to two antigens. A matrix of antigen—antibody complexes in a solution will then lead to a formation of a visible precipitate. In the precipitin test, a soluble antigen and antibody diffuse towards each other, and a visible precipitate forms when the two solutes meet at an optimal concentration [4]. Usually, a blood specimen is mixed with test antigen to detect patient antibodies, most often in suspected fungal infection or pyogenic meningitis.

Because a positive result requires a large amount of antibody or antigen, sensitivity is low.

To improve sensitivity and to decrease the amount of immunoreagents become possible to use as analytical signal not formation of precipitate, but more sensitive changing of light scattering through antigen–antibody immunocomplex formation. Application of nephelometers allowed quantitative measuring of the turbidity of solution and the method was called nephelometry.

There are many types of precipitation tests (eg, Ouchterlony double diffusion and counter-immunoelectrophoresis). The double immunodiffusion, or Ouchterlony gel diffusion, takes place in an agarose gel, with adjacent wells filled with an unknown soluble antigen and the known antibody solution. If antigen specific for the known antibody is present, the two components produce a visible precipitin band in the gel.

In counter-immunoelectrophoresis, electrical current is applied to speed up the migration of soluble antigens and antibodies and concentrated them in small part of solution. In basic pH, most microbial antigens have a net negative charge migrating towards the positively charged electrode. Antibody molecules are only weakly negatively charged or neutral under alkaline conditions, and hence, they do not migrate in the electrical field but are carried towards the cathode by the buffer ions – a phenomenon called electroendoosmosis [4]. When target antigen and antibody are present, they will meet and form a visible precipitin band. Compared to passive diffusion, the electrophoretically enhanced system is much faster, requiring less than an hour for completion. In the 1970s, counter-immunoelectrophoresis was widely used in clinical laboratories, but it has now been replaced by faster, more sensitive and user-friendly immunochemical tests for antigen detection.

3. AGGLUTINATION

Complex formation between antibodies with their matching antigens on a surface of large particles, such as animal cell, erythrocytes, or bacteria, results in antibodies cross-linking the cells-forming visible clumps. This reaction is termed as agglutination. Herbert Edward Durham and Max von Gruber discovered specific agglutination in 1896; the process became known as Gruber–Durham reaction. Von Gruber introduced the term agglutinin (from the Latin) for any substance that caused agglutination of cells.

First this reaction was used as the basis for a test for typhoid fever. French physician Fernand Widal found that blood serum from a typhoid carrier caused a culture of typhoid bacteria to clump, whereas serum from a typhoid-free person did not [5]. This test for typhoid fever was the first widely used example of serum diagnosis.

Another important practical application of the agglutination reaction was discovered by Austrian physician Karl Landsteiner. In 1900 he found out that the blood of two people under contact agglutinates, and established this effect was due to contact of blood with blood serum. Landsteiner discovered the ABO blood group system by mixing the red cells and serum of each of his staff. He demonstrated that the serum of some people agglutinated the red cells of other [6]. Landsteiner's discovery that red cells from some people are agglutinated by serum from others was the beginning of the haematological revolution. From these early experiments, he identified three types, called A, B and C (C was later to be renamed O for the German 'Ohne', meaning 'without', or 'Zero', 'null' in English). The fourth less frequent blood group AB, was discovered a year later. In his research Landsteiner found out that blood transfusion between persons with the same blood group did not lead to the destruction of blood cells, whereas this occurred between persons of different blood groups [7]. Landsteiner's agglutination tests and his discovery of ABO blood groups was the start of the science of blood transfusion and serology which has made transfusion possible and safer. In 1930, Landsteiner received the Nobel Prize in physiology and medicine for his work.

In modern agglutination tests (eg, latex agglutination and coaggregation), a particle (latex bead or bacterium) is coupled to a reagent antigen or antibody. The resulting particle complex is mixed with the specimen (eg, serum); if the target antibody or antigen is present in the specimen, it cross-links the particles, producing measurable agglutination. If results are positive, the body fluid is serially diluted and tested. Agglutination with more dilute solutions indicates higher concentrations of the target analyte (antigen or antibody). The titre is correctly reported as the reciprocal of the most dilute solution yielding agglutination; for example, 64 indicates that agglutination occurred in a solution diluted to 1/64 of the starting concentration. Agglutination is widely used as a method of identifying specific bacterial antigens, and in turn, the identity of bacteria. Because the clumping reaction occurs quickly and is easy to produce, agglutination is a common technique in diagnosis. Usually, agglutination tests are rapid but it sensitivity is less than many other methods (eg, label-based methods).

In current clinical medicine, agglutination reactions have many applications. They can be used to type blood cells for transfusion, for identification of bacterial cultures and to detect the presence of a specific antibody in the serum of the patient. Agglutination is primarily used to check if a patient has a bacterial infection or not.

The main difference between precipitation and agglutination pertains to the size of antigens. In the case of precipitation, antigens are soluble molecules while in the case of agglutination; antigens are large, insoluble molecules. Another difference between these two serological reactions is that agglutination reaction is more sensitive than precipitation reaction because a lot of soluble antigens and antibody molecules are required to form a visible precipitate. However, it is possible to make a precipitation more sensitive by converting it into agglutination reaction. This can be achieved by attaching soluble antigens to large, inert carriers such as erythrocytes or latex beads.

Different variants of agglutination were developed, such as complement fixation tests, haemagglutination inhibition assay, latex agglutination, staphylococcal agglutination and immunoturbidimetric assays, and part of them is still actual.

Complement fixation test. A variant of agglutination reaction is a, complement fixation tests, which are used in public health laboratories. The complement fixation test was first discovered by Jules Bordet and Octave Gengou in 1901 [8]. The complement fixation test is employed not only antigen–antibody interaction, but also the complement system. The complement system is a part of the innate immune system. It helps (or complements) to clear pathogens from an organism. The complement system consists of over 30 proteins and protein fragments normally circulating in the blood as inactive precursors. When stimulated by one of several triggers, proteases, in the system cleave specific proteins to release cytokines and initiate an amplifying cascade of further cleavages. The end result of this activation cascade is massive amplification of the response and activation of the cell-killing membrane attack complex. The complement system reacts with antigen–antibody complexes. If this reaction occurs on a cell surface, it will result in the formation of transmembrane pores and therefore destruction of the cell.

In 1894, Richard Pfeiffer, a German scientist, had discovered that when cholera bacteria were injected into the peritoneum of a guinea pig immunized against the infection, the pig would rapidly die. This bacteriolysis, Bordet discovered, did not occur when the bacteria was injected into a nonimmunized guinea pig, but did so when the same animal received the antiserum from an immunized animal. Moreover, the bacteriolysis did not take place when the bacteria and the antiserum were mixed in a test tube unless fresh antiserum was used.

However, when Bordet heated the antiserum to 55°C, it lost its power to kill bacteria. Finding that he could restore the bacteriolytic power of the antiserum if he added a little fresh serum from a nonimmunized animal, Bordet concluded that the bacteria-killing phenomenon was due to the combined action of two distinct substances: an antibody in the antiserum, which specifically acted against a particular kind of bacterium; and a nonspecific substance, sensitive to heat, found in all animal serums, which Bordet called 'alexine' (later named 'complement'). In a series of experiments conducted later, Bordet also learnt that injecting red blood cells from one animal species (rabbit cells in the initial experiments) into another species (guinea pigs) caused the serum of the second species to quickly destroy the red cells of the first. And although the serum lost its power to kill the red cells when heated to 55°C, its potency was restored when alexine was added. It became apparent to

Bordet that haemolytic (red cell destroying) serums acted exactly as bacteriolytic serums; thus, he had uncovered the basic mechanism by which animal bodies defend or immunize themselves against the invasion of foreign elements. Eventually, Bordet and his colleagues found a way to implement their discoveries and develop a method to visually determine the presence of bacteria in a patient's blood serum. This process became known as a complement fixation test [9].

At the beginning, complement has been considered to function only in conjunction with antierythrocyte and antibacterial antibodies, to affect the lysis or death of target cells [10]. This finding led Albert Neissner to show that complement fixation might be employed for the determination of proteins for forensic purposes and August Paul von Wasserman to show that this approach might also be employed for the diagnosis of syphilis and other infectious diseases [11].

The complement fixation test is a diagnostic technique that was used to detect the presence of infectious agents in the blood, including those that cause typhoid, tuberculosis and, most notably, syphilis (the Wassermann test). The complement fixation test is used to detect the presence of either specific antibody (more often) or specific antigen. It was widely used to diagnose infections, particularly with microbes that are not easily detected by culture methods, and in rheumatic diseases.

The basic steps of a complement fixation test are as follows.

The first step is a negation of any effects of different level of amount of complement proteins in patient's serum. Thus, the complement proteins in the patient's serum must be destroyed and replaced by a known amount of standardized complement proteins. To standardize complement fixation test, a patient's serum is heated in such a way that all of the complement proteins — but none of the antibodies — within it are destroyed. This is possible because complement proteins are much more susceptible to destruction by heat than antibodies. Then a known amount of standard complement proteins (usually obtained from guinea pig serum) are added to the serum. Also the antigen of interest is added to the serum.

On the second step, to determine whether complement has been fixed, sheep red blood cells and antibodies against sheep red blood cells are added. If patient's serum contains antibody against the antigen of interest, the complement is already attached to the antigen—antibody complex in the serum and not able to induce sheep red blood cells lysis because of no complement left in the serum. Therefore, these cells settle to the bottom of the tube. If patient's serum does not contain antibody against the antigen of interest, complement is free to react with complex sheep red blood cells—antibody. As the complement is not depleted it will react with the sheep red blood cells—antibody complexes, lysing the sheep red blood cells and spilling their contents into the solution, thereby turning the solution pink.

While detection of antibodies is the more common test format, it is also possible to detect the presence of antigen. In this case, the patient's serum is supplemented with specific antibody to induce formation of complexes; addition of complement and indicator sheep red blood cells is performed as described in the preceding paragraphs.

The complement fixation test was widely used to diagnose infections, particularly with microbes that are not easily detected by culture methods, and in rheumatic diseases. Complement fixation test meets the following criteria:

1. it is convenient and rapid to perform;
2. the demand on equipment and reagents is small;
3. a large variety of test antigens are readily available.

Although complement fixation test is considered to be a relatively simple test, it is a very exacting procedure because variables are involved. In clinical diagnostic laboratories it has been largely superseded by other methods such as enzyme-linked immunoassay and by DNA-based methods of pathogen detection, particularly polymerase chain reaction.

Haemagglutination. The haemagglutination (from the Greek *haima* – blood and Latin *agglutinatio* – binding) is an agglutination of red blood cells in the presence of bacteria, viruses, toxins, and so on, which adsorb on the surface of red blood cells. As each of the agglutinating subjects attaches to multiple red blood cells, a lattice-structure will form. The haemagglutination assay and its extension, the haemagglutination inhibition assay, were invented by American virologist George Hirst. In 1941, Hirst discovered that adding influenza virus particles to red blood cells caused them to agglutinate or stick together forming a lattice, a phenomenon called haemagglutination [12]. Hirst developed this reaction into the haemagglutination assay, which allows the amount of virus in the sample to be measured. He soon realized that the haemagglutination assay could easily be adapted to measure the levels of antibody specific to the virus strain in human serum: any antibodies present bind to the influenza virus particles, prevent them from crosslinking red blood cells and so inhibit haemagglutination [13].

Haemagglutination inhibition assay is an indirect or passive agglutination assay where the antigen is bound to an inert substance and then mixed with patient serum containing the possible antibody. Serially diluted serum samples are placed in a V-well microtitre plate to which the viral antigen and the red blood cells are added. The last serial dilution that yields total inhibition of agglutination is the serum titre for the patient. This haemagglutination inhibition assay can be applied to many viruses carrying a haemagglutinin molecule, including rubella, measles, mumps, parainfluenza, adenoviruses, polyomaviruses and arboviruses, and is still widely used in influenza surveillance and vaccine testing. Haemagglutination inhibition assays are quite tedious and labour intensive because they require fresh erythrocytes and antigen dilutions each time the test is performed. At the moment, one of three immunochemical treponemal tests used for the identification of syphilis is based on a qualitative haemagglutination test. In this assay, anti–*Treponema pallidum* antibodies are detected using sheep red blood cells as a carrier for the antigen. For the test, the patient serum is serially diluted and added to both *T. pallidum*–sensitized and unsensitized sheep red blood cells, and the results are reported as titres for the haemagglutinization reaction [4].

Latex agglutination test. The first description of a test based in latex agglutination was the rheumatoid factor test proposed by Jacques Singer and Charles Plotz [14] in 1956. Since then, tests to detect microbial and viral infections, autoimmune diseases, hormones, drugs and serum proteins have been developed and marketed by many companies worldwide [15]. Latex agglutination tests become very popular in clinical laboratories. These tests have been applied to the detection of over 100 infectious diseases, and many other applications are currently available.

In latex agglutination procedures, an antibody (or antigen) coats the surface of latex particles (sensitized latex). When a sample containing the specific antigen (or antibody) is mixed with the milky-appearing sensitized latex, it causes visible agglutination. The degree of agglutination plotted as a function of agglutinant concentration follows a bell–shaped curve similar to the precipitin one. Latex particles are used to magnify the antigen–antibody complex [15].

At the beginning many of the latex agglutination tests developed were performed manually and the agglutination was detected by visual observation. Although quite useful in the laboratory and cheap due to the absence of equipment needs, these manual assays suffered from lack of consistency in end-point readouts. It had been established that about 100 clumps must be seen to determine agglutination, and that these clumps must be of about 50 μm in size to be seen by eye [16]. Since the most common size for latex in these tests is 0.8 μm, about 10^5 latex particles would be required to make one visible aggregate, and about 10^7 particles would be needed to determine agglutination in a given test. Based on these calculations and assuming that about 10 bonds are required per particle to hold them together, Bangs [16] has evaluated the sensitivity of such a manual tests to be in the order of picograms.

Latex agglutination tests are prone to false-positive results due to nonspecific agglutination reactions. Hence, it is important to include proper positive and negative controls in the test. Sample pretreatment methods, like centrifugation, boiling, or filtration can be helpful in elimination or minimization of nonspecific agglutination. Some specimens, like urine, can be concentrated by centrifugation or membrane filtration to increase the test sensitivity. Latex agglutination test kits are commercially available. The tests are easy to perform and sensitive. However, they might not be the fastest method of choice [4].

During the years, several approaches to detect latex particles agglutination have been described using spectrophotometers and nephelometers which measure absorbed or scattered light, in place of the human eye. Agglutination detects in tubes or 96-well plates [16]. Some companies have devised particle counters appropriate to detect very small clumps. Angular anisotropy or quasi-elastic light scattering have also been applied to latex agglutination tests. Any of these new approaches allows to increase sensitivity and to improve standardization, and the procedures involved may be automated [15]. In staphylococcal coagglutination tests, intact formalin-killed *Staphylococcus aureus* cells were used for visualization instead of latex beads [17]. The cell walls of these organisms contain protein A, which binds to the Fc portion of the IgG antibody, leaving the Fab portion available to react with specific antigens.

Coagglutination is more susceptible for nonspecific reactions, and sample preparation is even more important than when using latex particles. Coagglutination is highly specific, though, and it is often used to confirm the identification of bacterial colonies in culture plates, but it is often not the best choice for rapid antigen detection in a clinical specimen. Even though latex agglutination tests are rapidly being replaced by faster methods like lateral-flow test strips, latex agglutination tests are still routinely used for culture confirmation and serotyping in many clinical laboratories [4].

4. RADIOIMMUNOASSAY

Discovery and possibilities. The next important era in clinical tests development began with the application of the labels for 'visualization' of an antigen–antibody interaction. The first label was radioactive and was introduced by Solomon Berson and Rosalyn S. Yalow in 1959 [18] for insulin assay. Berson and Yalow recognized the demands for sensitivity and specificity of an assay system and introduced radiometric methods to measure the antigen–antibody reaction. Radioimmunoassay (RIA) become a powerful tool for years and introduced a lot of theoretical and technical breakthrough in immunoassay practice. The important outcome of these experiments is not only first introduction of label-based assay, but also competitive assay format. This immunoassay is based on the competition between labelled antigen (tracer) and unlabelled antigen for specific antibody sites, forming antigen–antibody complexes.

In 1955 Berson and Yalow, injected two groups of human subjects with ^{131}I-labelled insulin: One group had previously been treated with insulin, while the second group had never received exogenous insulin. The disappearance of ^{131}I-insulin measured by trichloroacetic acid precipitation in serially obtained blood samples was determined. Insulin disappearance from the blood of patients who had received insulin previously was markedly slower than in the patients who had never received insulin. Electrophoretic separation demonstrated that ^{131}I-insulin was bound to a globulin fraction in the first group of patients. The extraordinary specificity of this reaction and its quantitation were demonstrated and made a basis of the new assay [19,20]. Aside from making this conceptual breakthrough, they also prepared and evaluated antisera with satisfactory specificity and with high-affinity constants capable of detecting physiological concentrations of unique peptides, as well as iodinating peptides at high specific activity with minimal denaturation. This was essential for the quantitation of the sensitive antigen—antibody affinity constants, which were much greater than immunologists had previously supposed. Concluding that at these concentrations precipitation of the complexes was unlikely, they adopted and applied methods for the separation of antibody-bound radioactivity and the free radioactive (hormonal) antigen [21].

In addition to providing an insight into insulin secretory physiology in health and disease, these investigations stimulated the rapid appreciation of the technical subtleties in the handling, storage, interaction and separation of dilute labelled and unlabelled peptides and antisera [22,23].

Despite the significance of the RIA of plasma insulin, the general applicability of the method remained to be demonstrated. The development of the RIA of human growth hormone accomplished this [24,25]. Thereafter it was apparent that although each hormone (or another analyte) might vary in antigenicity and have its own peculiar physical properties influencing solubility and iodination and affecting the binding to surfaces, a method had been developed to study and pursue the answers to these problems [21].

The role of RIA in patient care, nuclear medicine, blood banking, allergology, endocrinology clinical chemistry, and clinical and basic medical investigation has been well documented in several reviews [21,26—29].

Principle of RIA. To perform RIA, a known amount of a radioactive-labelled antigen (tracer) is mixed with a known amount of antibody specific for this antigen, and as a result, the two form immunocomplex, specifically binding to one another. Then, a sample containing an unknown quantity of the antigen is added. This causes the unlabelled (so-called 'cold') antigen from the sample to compete with the radiolabelled antigen ('hot') for restricted amount of antibody-binding sites. As the concentration of 'cold' antigen is increased, more of it binds to the antibody, displacing the tracer, and reducing the ratio of antibody-bound radiolabelled antigen to free radiolabelled antigen. Separation of the antibody—antigen complexes from free antigen is achieved by precipitation of the antibody-bound tracer with either a secondary antibody solution directed against the genus- or species-specific immunoglobulins of the primary antibody, or by use of polyethylene glycol. Then the bound antigens are separated from the unbound ones, and the radioactivity of the free antigen, not bound to the immunocomplex with antibody, remaining in the supernatant is measured using a gamma counter. This assay is performed in homogeneous format, and it was developed first. To develop the first RIA insulin antibodies were purified from insulin-treated subjects and used. In order to detect native insulin from blood samples, tyrosine residues of insulin were labelled with radioactive iodine (^{131}I), which competed with unlabelled insulin from the sample for binding to the antibody. After an incubation period, the bound and free fractions were separated from each other by paper chromatography, and the radioactivity values of the separated fractions were measured by a scintillation counter [29,30].

In heterogeneous format antibody is attached to the solid support (an insoluble material). Limited amount of antibody-binding sites on the surface of solid support can either bind unlabelled 'cold' ligand or radiolabelled 'hot' ligand. As the amount of analyte increases, there is consequently less radiolabelled ligand bound. The analyte can come from either a 'calibration standard' or from the sample. After incubation is performed, solid support is finally washed out with water or buffer solution and counted for quantitation of the bound tracer.

In 1968, Miles and Hales published their first results of an 'immunoradiometric' technique with radioactive-labelled antibodies rather than labelled antigen for measuring insulin in human plasma [31]. These labelled antibodies were used in other kind of heterogeneous format. An analyte (eg, suspected allergen) was bound to the solid support and the sample (patient's serum) is added. If the serum contains antibodies to the allergen (IgE is the antibody, associated with Type I allergic response), those antibodies will bind to the attached allergen. Radiolabelled antihuman IgE antibody is added where it binds to those IgE antibodies already bound to the solid support. The unbound antihuman IgE antibodies are washed away. The amount of radioactivity is proportional to the serum IgE for the allergen.

The type of solid supports is especially important from the point of the development of rapid tests. Heterogeneous format gives possibility for preconcentration of the analyte (or analytes) on the solid support and fast simple washing steps. As RIA was the first label-based heterogeneous method, many types of possible materials and forms of solid supports are introduce in the practice. A lot of water-insoluble polymers of different chemical nature were tested and as optimal either poly-(tetrafluoroethylene-g-isothiocyanatostyrene) [32] or Sephadex−isothiocyanate [33] were found. After that the former polymer has been prepared in the form of small discs [34].

During examination of various polymeric materials for applicability to solid-phase RIA, it became apparent that certain unsubstituted polymers may adsorb antibody that can then bind an adequate quantity of radioactive tracer antigen for use in the assay. In contrast, adsorption of antibody to glass was negligible [35]. The two most commonly available disposable plastic tubes were manufactured from polypropylene or polystyrene, both of which gave satisfactory results in the assay. The assay was performed in tubes suitable for use in a gamma counter, adequate uniformity of replicates was achieved, and satisfactory results were obtained by overnight incubation without initial disequilibration. Expense was minimized by use of the same tube for all phases of the assay procedure; thus incubation, separation of bound and free tracer, and counting of the bound tracer were all performed in a tube costing only a few cents. The advantages of use of antibody-coated tubes were currently being incorporated into a system for automated RIA [35]. The next step of simplification and automatization was application of microtitre well plates [36] with related microplate scintillation counter. Simple paper disc were used for RIA by [37].

Radioactive labels. Radioactive labels were the first group of labels providing high sensitivity. Their radioactivity (analytical signal) was measured with liquid scintillation counter or gamma counter.

The specific activity of the labelled compound is an important consideration, since the sensitivity of the assay is strongly dependent on the total amount of the tracer used. Clearly, low specific—activity tracers will have a greater molar concentration of tracer per radioactive unit and, consequently, less sensitivity. Also label should not occur with great frequency in organic compounds, such as, for example, of carbon atoms and ^{14}C isotope. Tritium has generally been limited to use in steroid hormone and drug assays. For analytes with the lower concentration, such as peptide hormones in plasma, high specific—activity labelled compounds are needed. This requirement has limited the radionuclides useful for this purpose to ^{131}I and ^{125}I These isotopes were attached to the tracer molecules through the tyrosyl molecule. One or two iodine atoms may react with each tyrosyl moiety. The total number of tyrosyls per tracer molecule determines the maximal capacity of the molecule for iodine. It is desirable from the point of view of maximal specific activity to incorporate as many radioiodine atoms per molecule as possible. However, an increase in iodine content places a practical limitation through loss of immunological cross-reactivity of the iodinated antigen. Thus, it was necessary to accept a compromise, sacrificing maximal activity for other considerations. Since potential loss of immunerecognition is the price paid for the introduction of each iodine atom per tracer molecule (except for naturally iodinated molecules, such as triiodothyronine and thyroxine), it is worthwhile to attempt to incorporate only radioiodine through the use of pure preparations of the radionuclide selected. While pure ^{131}I has a theoretical potential-specific activity, about 7 times higher than the specific activity of pure ^{125}I, the greater isotopic abundance of ^{125}I, has resulted in ^{125}I emerging as the radionuclide of choice for use in RIA systems [21]. The longer half-life of ^{125}I (lifetime 56 days), compared to ^{131}I (lifetime 8 days), while theoretically a disadvantage, becomes an advantage in the real world of laboratory operation. For equivalent amounts of radioactivity, the labelled substrate concentration increases with radionuclidic decay, so that ^{131}I-labelled antigen per tube in an assay performed 8 days after iodination will contain twice the amount of antigen as it would have contained on the day of preparation. By contrast, ^{125}I-labelled compounds, with a 60-day physical half-life, will have 91% of their original specific activity after 8 days. It should be recalled that in the original studies ^{131}I (β and γ radiation) was used for the labelling because no alternatives were available at that time. The potential health problems related to the use of radioactive materials were greatly diminished when manufacturers such as Amersham and NEN began marketing ^{125}I (weak γ radiation) preparations of sufficiently high specific activity and purity [38].

Application. Insulin was the first analyte to measure, followed with the growth hormone detection. The RIA for insulin has paved the way for the development of immunoassays for thousands of other analytes over the ensuing half a century. Despite the significance of the RIA of plasma insulin, the general applicability of the method remained to be demonstrated. The development of the RIA of human growth hormone accomplished this [24,25]. Thereafter, it was apparent that although each hormone (or allergen, or another analyte) might vary in antigenicity and have its own peculiar physical properties influencing solubility and iodination and affecting the binding to surfaces (for heterogeneous assay), a method had been developed to study and pursue the answers to these problems [21]. The RIA technique immediately caught the imagination of many researchers and clinicians, and RIAs for new analytes were published at a rapid pace and variants of the method were rapidly developed [38].

Table 1 presents list of the compounds assayed by RIA. RIA can be applied to most forms of biological material, such as blood, urine and tissue preparations. It was possible to automate at least parts of the assay procedures, and calculation of the results by computer facilitates the handling of large numbers of samples simultaneously.

With the inclusion of a separation step several analytes (eg, antibodies related to allergic response) could be measured simultaneously in small sample volumes. The required plasma volumes are also small enough to allow for frequent sampling in the same patient when, for example, hormone profiles and rapid fluctuations should be measured. Very fast RIA almost completely conquered the field of hormone analysis and has made obsolete almost all previously used techniques for routine analysis. The assays offered satisfactory reliability, and the sensitivity was sufficient for many medical problems. It was possible to detect as low as few picograms of analyte in the experimental tube when using antibodies of high affinity ($K_d = 10^{-8}$–10^{-11} M).

A decision by Drs Berson and Yalow to not patent this technology accelerated the progress of RIA and the full area of immunoassay development.

Disadvantages. Despite an important role of RIA in clinical assay for wide range of analytes, patient care, nuclear medicine, blood banking, allergology and endocrinology, for now, it is almost not possible to meet RIA in clinical laboratories. While conceptual breakthrough of RIA technique for different areas of assay and medicine, which impossible overestimate, there are some principal unavoidable disadvantages:

- Although the RIA technique is extremely sensitive and very specific, requiring specialized equipment, it remains among the least expensive

TABLE 1 Radioimmunoassay Application

	Class of Compounds	Compounds
Hormones	Steroid hormones [21,39]	Oestrogens (estrone, estradiol, estriol, estrone-sulphate, 2-hydroxyestrone, 15 α-hydroxyestriol) Corticosteroids (cortisol, corticosterone, deoxycorticosterone, aldosterone) Androgens (testosterone, dihydrotestosterone, testosterone-glucuronoside, dehydroepiandrosterone, dehydroepiandrosterone-sulphate, androstenedione, androstene-3β, 17β-diol androsterone) Progestins (progesterone, 17-hydroxyprogesterone, pregnenolone, 17-hydroxypregnenolone) Synthetic steroids (medroxyprogesterone, dexamethasone, prednisone, cortisone, hydroxycortisone)
	Peptide hormones [21]	Growth hormone, adrenocorticotropi c hormone (ACTH), melanocyte-stimulating hormone (MSH), thyrotropin-releasing factor (TRF), thyroid-stimulating hormone (TSH), follicle-stimulating hormone (FSH), luteinizing hormone (LM), prolactin, lipotropin (LPH), vasopressin, oxytocin, human chorionic gonadotropin (HCG), human chorionic somatomammotropin (HCS), insulin, prolnsulin, C-peptide, glucagon, enteroglucagon, parathyroid hormone (PTH), calcitonin (CT), gastrin, secretin, cholecystokinin (CCK), angiotensins, bradykinins, somatomedins
Drugs [28]	Analgesics	Aspirin, codeine, etorphine, fentany 1, hydromorphone, methadone, morphine, pentazocine
	Antibiotics	Adriamycin, amikacin, chloramphenicol, daunomycin, gentamicin, isoniazid, penicillins Sulphonamides, tetracycline
	Anticonvulsants	Phenytoin
	Antidiabetic agents	Sulthiame
	Antihypertensives	Glibenclamide

TABLE 1 Radioimmunoassay Application—cont'd

	Class of Compounds	Compounds
	Antineoplastic agents	Saralasin
	Barbiturates	Methotrexate, barbitone, pentobarbitone
	Cardiac depressants	Procaineamide
	Cardiac glycosides	Digitoxin, digoxin, gitaloxin, gitoxin, ouabain
	Hallucinogens	Cannabinoids, lysergide, mescaline, 2,5-dimethoxy-4-methylamphetamine
	Muscle relaxants	Tubocurarine
Allergenes		IgE [40] and IgE antibodies [41,42]
		Antigen P1 [43]
Insecticides [28]		Aldrin, DDT, dieldrin, malathion
Tumour-associated antigens [44,45]		
Histamine [46]		

methods to perform such measurements. The cost of the equipment is the limiting factor for RIA application.

- Because radioactivity poses a potential health threat, RIA application in laboratories requires special precautions and licensing, since radioactive substances are used.
- The counting of radioactivity is time consuming and the labels are short living.
- In addition, the radioactive waste problem, the requirements of building special laboratory facilities are among limiting factors of this technique.

RIA Outlook. RIA discovering, development and application was a bright page in history of immunoassay, rapid tests and clinical assay development, and initiated a lot of theoretical and practical research and discoveries in related areas. RIA disadvantages had very positive role: they promote development of new things, in the case of immunoassay they promote discovering of new labels solid supports and as a results new formats.

Therefore alternative methods were awaited with great interest. Of the further developments, enzyme immunoassay (EIA), where the radiolabel is replaced by an enzyme, appear to overcome some of the problems of RIA. Since in these assays analytical signal will not be the radioactivity but optical density, fluorescence or chemiluminescence intensity and enzyme-immunoassay could be introduced into all clinical chemistry laboratories which had been prevented from using RIA because of the high costs of counting equipment.

5. ENZYME IMMUNOASSAY

The first published EIA and enzyme-linked immunosorbent assay (ELISA) systems differed in assay design, but both techniques are based on the principle of immunoassay with an enzyme rather than radioisotope as the reporter label. Two scientific research groups independently and simultaneously developed this idea in 1971 and executed the necessary experiments to demonstrate its feasibility. The ELISA technique was conceptualized and developed by Peter Perlmann and Eva Engvall at Stockholm University, Sweden [47] who described a method for the analysis of the concentration of immunoglobulin G in serum with alkaline phosphatase as the reporter label. The EIA technique was introduced by Anton Schuurs and Bauke van Weemen at the Research Laboratories of NV Organon, Oss, the Netherlands [48] as pregnancy test.

At scientific meetings of the European RadioImmunoAssay Club in Basel in the early 1970s, the idea of using enzyme labels was met with scepticism and incredulity: How could so bulky and large molecule as an enzyme be attached to an antigen or antibody without sterically hindering the immunochemical antigen–antibody binding [38]? This objection on principle was nullified by carefully planned and executed experiments to demonstrate the feasibility of enzyme assays. Initial results were encouraging, and later the resounding success of the enzyme-(linked) immunoassay technique proved all sceptics wrong. Labelling of antigen with a suitable enzyme instead of isotope would allow to work without restrictions and risks related to radioactivity, and also offer certain advantages:

- enzyme–antigen conjugates can be stabilized, so that one preparation can be used for a long time;
- simpler equipment is usually required to measure enzyme activities than to measure radioactivity.

As prehistory of application enzymes as label for immunoassay, between 1966 and 1969, the group in Villejuif reported their successful results of coupling antigens or antibodies with enzymes such as alkaline phosphatase, glucose oxidase and others [49,50]. Avrameas and colleagues developed the coupling of these molecules by means of glutaraldehyde. Their purpose was to use the enzyme-labelled antigens and antibodies to detect antibodies or antigens by immunofluorescence, and they applied their tools to histopathology for histochemical detection antigens in tissue sections.

Developed by Avrameas and Uriel technique of coupling enzyme to antibodies was used for the first time by Engvall and Perlmann for the quantitative assay of antigens and antibodies. Immunoglobulin G (IgG) from rabbit was used as antigen. A sheep–antirabbit IgG was conjugated to BrCN-activated microcrystalline cellulose. Alkaline phosphatase from calf intestinal mucosa was conjugated to the rabbit IgG by use of glutardialdehyde method [50]. The conjugation experiments showed only 30–40% of the enzyme activity was lost during conjugation. ELISA showed similarly with RIA sensitivity and precision. Repeated determinations (6 times during 5 months) of one batch of IgG, diluted to give 49% inhibition in ELISA, gave a standard deviation of 1.5%. Conjugates made with alkaline phosphatase showed to be very stable and if stored in the cold they can be used for at least 6 months without any detectable loss of activity [47]. According to format, the technique [47] was the development of the heterogeneous radioimmunosorbent techniques for quantitative determination of antigen [33] use is made of radioactively labelled antigen and insolubilized antibodies, covalently coupled to cellulose or Sephadex.

Engvall's group applied the ELISA measurement tool to parasitology (eg, malaria [51] and trichinosis [52]), microbiology [53] and oncology [54]. It is interesting to mention, in the beginning of development ELISA-devoted articles called method as serological [51]. Gaining popularity and recognition the abbreviation ELISA becomes common in the titles without deciphering abbreviations.

Van Weemen and Schuurs published their innovative work on ELISA for quantification of human chorionic gonadotropin (HCG, 'pregnancy' hormone) concentrations in urine. Diazotized and CNBr-activated cellulose was used as adsorbents for antibody. As the reporter label they used the enzyme horseradish peroxidase (HRP), the most popular label for ELISA since then. Authors used two competitive EIA formats: assay labelled with HRP antigen competing with antigen for restricted amount of binding sites attached to cellulose-specific antibodies and variant of assay on the first step analyte from sample formed immunocompex with specific antibody, then labelled with HRP antigen was added and competes with analyte for restricted amount of binding sites of specific antibody in solution and finally mixture was added to cellulose microparticles with attached secondary antispices antibody to bind specific antibody [48].

It is important to mention, enzyme immunoassay with application of secondary antibody was found to be more sensitive and reproducible. Better sensitivity was explained by the assumption that some of the antibodies on the surface of adsorbents are accessible for unbound analyte–only HCG, but not for the high-molecular-weight conjugates with HRP, while others are accessible for all reaction partners. Excess of adsorbent with attached specific antibody impairs the sensitivity of the test system while excess of adsorbent with attached secondary antibodies does not. Better reproducibility became possible because of specific antibody solution can be added more accurately than a cellulose suspension. The amount of bound with secondary antibody cellulose in is not critical, since it has to be used in excess. The applicability of the EIA was illustrated by the determination of HCG in urines from 15 pregnant and 15 nonpregnant women, adjusted to a sensitivity level of 0.4 IU/mL, and by a haemagglutination inhibition assay with a sensitivity of 1 IU/mL. The correlation between these test methods was excellent. The HCG concentrations in all nonpregnancy urines were below the detection of either test [48].

The most interesting applications were initiated in the fields of bacteriology/parasitology/virology. Researches realized that the simplicity of ELISA compared very favourably with the cumbersome technology that was used in the field at that time (haemagglutination inhibition and complement fixation) and quickly developed a wide range of tests for various microorganisms. These gained almost immediate acceptance and greatly helped in making reliable infectious disease testing. Until then, such testing had been limited to only a small number of highly specialized laboratories [55].

The adoption of ELISA in fields such as endocrinology and oncology, in which RIA already had gained a strong foothold, was much slower than in infectious disease testing. The precision and sensitivity of ELISA lagged behind those of the well-developed and established RIA, and laboratories that were equipped for and accustomed to working with radioactivity did not see advantages in changing to work with nonisotopic methods. But possibility of automation has been a driving force that changed all these. While automation of methods involving radioisotopes created many problems, ELISA in couple of decades offered opportunities for fully automated random-access immunoassay systems, which diagnostics manufacturers fully exploited during the 1980s [55]. This led to more reliable testing at lower cost and facilitated the move of immunochemistry from specialized radioisotope departments into the general chemistry laboratory; ELISA thus also found its way into areas other than just clinical assay, such as food and feed control and further to environmental control.

The first wave of immunoassay automation was sparked in 1975 when Alister Voller and Dennis Bidwell initiated the use of 96-well microplates, also referred to as multiwell plates and microtitre plates for performing ELISA [56]. The earliest microplates were handmade and date back to 1954/55 in Hungary, where the need arose because of a severe influenza outbreak. Dr Gyola Takatsy wanted to speed up the diagnostic testing of patients and the common test tube was too cumbersome and slow. He machined six rows of 12 'wells'. However, common usage of the microplate began in the late 1950s when John Liner in United States had introduced a moulded version. The word 'microtitre' is a registered trademark of Cooke Engineering Company, and

Thermo Electron OY is the last listed owner of the trademark (US Trademark 72,128,338.) It is now more usual to use the generic term 'microplate'.

A microplate or multiwell is a flat plate/tray/panel/dish with multiple test tubes. A microplate typically has 6, 24, 96, 384 or even 1536 sample wells arranged in a 2:3 rectangular matrix. Some microplates have even been manufactured with 3456 or even 9600 wells, which brings microplates into nanorange and picorange.

Each well of a microplate typically holds between tens of nanolitres to several 100 millilitres of liquid. Microplates with an embedded layer of filter material were developed in the early 1980s by several companies, and today, there are microplates for just about every application in life science research which involves filtration, separation, optical detection, storage, reaction mixing or cell culture. Introduction of microplates onto market stimulates ELISA test kits production and commercialization in which either an antigen or an antibody is noncovalently bound to a solid-phase support. The main dates of microplate evolution are presented in Table 2.

The new tests become commercially successful late 1970s and early 1980s, when they matched the exquisite sensitivity of existing RIA systems for the same analytes. Technical advances led to automated pipetting devices (Micromedics; Hamilton), multichannel pipettes (Lab Systems), and microtitre plate readers and washers, and in the 1980s fully automated test instruments were manufactured by Boehringer-Mannheim and Abbott, among others. Such automated systems have come to stay in laboratories [38].

In the early 1970s, blood-bank screening for virological diseases such as hepatitis B antigen was done either by (semi)automated RIA or nonradioactive but rather cumbersome haemagglutination tests. In 1976, Organon Teknika developed and marketed a highly successful ELISA system for the hepatitis B surface antigen [57], featuring a 96-well microtitre plate format. This test became the first commercially available ELISA. Other microbiological and virological diagnostic tests soon followed, for example, for hepatitis B antigens, rubella antibodies, toxoplasma antibodies, and in the 1980s, an ELISA system for detection of human immunodeficiency virus antibodies.

Historically first and most popular detection was photometrical, based on the colour change in the presence of chromogenic substrate. Traditional ELISA typically involves chromogenic substrates that produce some kind of observable colour change to indicate the presence of antigen or analyte. When enzymes (such as peroxidase) react with appropriate substrates (such as 3,3′,5,5′-tetramethylbenzidine), a change in colour occurs, which is used as an analytical signal. Newer ELISA techniques use fluorescent, chemiluminescent, electrochemiluminescent and other kind of signals, which is produced in the presence of corresponding substrates. These new reporters have sufficiently higher sensitivities.

At current days, the number of analytical and clinical investigations relying on ELISA-based and ELISA-related techniques worldwide is exceedingly large. The

TABLE 2 The Main Dates of Microplate Application Evolution

1951	Dr. Gyula Takatsy created the first microplate by constructing six rows of 12 wells into a block of acrylic, which were used in place of test tubes and the first form of a microplate automation tool, a loop that mixed and transferred a predefined volume from one well to another, used in serial dilution testing.
1953	The first manufacturer, the American Linbro Company, began mass producing the moulded vacuum formed 96-well microplates.
1964	Dr Sever and Cooke Engineering introduced the first manufactured screw machines loops and droppers called the Microtiter. Lab technicians held between 8 and 12 loops in their hands while twirling them and moving them from row to row in a plate.
1967	Tom Astle produced a fully automated serial dilution instrument, called the Autotiter.
1974	Centers for Disease Control (CDC) in London began using microplates for ELISA diagnostics and quality control techniques.
1976	The demand for advanced instruments to perform ELISAs prompted the manufacturer lab systems, now part of Thermo Fisher Scientific, to evolve the early microplate readers into the Multiskan photometer.
1990	Roy Manns introduced first 96-well microplate with clear bottom
1992	Continuing advancements in various sized microplates to find the right combination of thermal mass and capacity, led to Genetix producing the first 384-well plate.
1992	First commercial 384-well plate
1992	View plate 96 for luminescence
1996	First 1536-well plate
1997	BMG LABTECH introduced the first multidetection microplate reader that included fluorescence polarization with the BMG POLARstar.
1998	The standardization of microplates led to an increased ability to develop automation equipment to move, sort and wash plates in future instruments.
2009	BioTek combines both a microplate washer and reagent dispenser in one instrument

Data from http://www.labmanager.com.

numbers of measurements and determinations using immunoassay for routine patient care are astronomical. The clinical impact of ELISA as nonradioactive variants of immunoassays is indeed overwhelming. Among solid supports applicable for ELISA, porous membranes allowed to combine antigen–antibody interaction with their movement through volume of material. This promotes creation of several new trends and groups of immunochemical methods.

6. WESTERN BLOTTING

The history with two groups independently developed and published EIA in 1971 repeated in 1979, then three groups developed and described new analytical technique for detection of specific proteins, called western blot. That year group of George Stark, Stanford University, published the paper that described the transfer of proteins by capillary action from a polyacrylamide/agarose gel with or without the denaturing agent sodium dodecylsulphate onto a special membrane [58]. This group was already famous for developing the RNA blotting technique known as 'northern blotting' [59].

Harry Towbin, the Friedrich Miescher Institute for Biomedical Research in Switzerland, was trying to figure out the specificity of antibodies against proteins in complex macromolecular structures. Along with Theophil Staehelin at Roche, he began to work out a method that would allow them to establish which antibody bound to which component of the ribosomal complex [60]. By that point, DNA and RNA blotting methods were popular, so the idea of transferring proteins out of a gel and onto a membrane was natural.

W. Neal Burnette, the Fred Hutchinson Cancer Research Center, Seattle, submitted his manuscript also in 1979, but it was rejected, then eventually published in 1981 [61]. He developed his technique independently, including the electrophoretic transfer step, but became aware of Stark's and Towbin's publications before he submitted his in 1979. His procedure was finally published in 1981 after it had already become famous through distribution of preprints to colleagues and friends. Burnette definitely gave the technique the name 'western blotting'. The name was a joke based on the DNA blotting technique called 'Southern blotting', which was named after its inventor, Edwin Southern [62]. Detection of RNA is termed 'northern blot' by George Stark. The laboratory of Burnette was on the west coast, so the nickname 'western blot' was given to the technique and stick to it.

Now the western blot is a widely used analytical technique used to detect specific proteins in a sample of tissue homogenate or extract both with routine and exploring purposes. It uses gel electrophoresis to separate native proteins by 3D structure or denatured proteins by the length of the polypeptide. The proteins are then transferred to a membrane, where they are stained with specific to the target protein antibodies. The gel electrophoresis step is included in western blot analysis to resolve the issue of the cross-reactivity of antibodies. For visualization of the immunocomplex protein-specific antibody radiolabels and then enzymes and nanoparticles were used.

7. LATERAL-FLOW IMMUNOASSAY

A well-known and very popular application of EIA is immunochromatography in which a capture antibody is immobilized onto a surface of a porous membrane, and a sample passes along the membrane. Analyte in the sample is bound by the antibody which is coupled to the detector reagent. As the sample passes over the zone to which the capture reagent has been immobilized, the analyte detector reagent complex is trapped, and a colour develops in proportion to the analyte present in the sample.

The main application driving the early development of the solid-phase, rapid-test technology was the human pregnancy test, which represented continual historical interest in urine testing for medical diagnostic purposes. This particular testing application made great strides in the 1970s, as a result of improvements in antibody generation technologies and significant gains in understanding of the biology and detection of HCG, derived largely from the work performed by Judith Vaitukaitis and coworkers [63].

To fully develop the lateral-flow test platform, a variety of other enabling technologies are required. These include technologies as diverse as antibody generation, nitrocellulose membrane manufacturing, fluid dispensing and processing equipment, as well as development and manufacturing methodologies. Many of these facilitative technologies had evolved throughout the early 1990s, to the point where many are now mature, off-the shelf technologies [63].

One of important steps to development of modern lateral-flow immunoassay (LFIA) systems was introduction of sol particles as labels. Colloidal gold particles, on which antibodies are adsorbed, have been used as electron-dense markers for electron microscopical investigation of cell surfaces [64]. Leuvering et al. were first to describe in 1980 the use of inorganic (metal) colloidal particles as a label for immunoassays. This group using microtitre-plate technique built dose—response sandwich immunoassays curves for human placental lactogen and HCG, using conjugates consisting of antibody-coated colloidal gold or silver particles. Several techniques were used to measure the amount of bound conjugate, naked eye, colorimetry and carbon rod atomic absorption spectrophotometry. At higher antigen concentrations the results of the assay could be read by the naked eye. Using gold particles as label and atomic absorption spectrophotometry as detection method, they found a detection limit for a sandwich sol particle immunoassay was equal to that of an optimized competitive RIA. Obtained detection limits are presented in Table 3 [65].

In this pioneer work also was found the possibility for simultaneous determination human placental lactogen and HCG, using microtitration plates, coated with a mixture of anti-HPL and anti-HCG, and a mixture of silver particle anti-HPL conjugate and gold particle anti-HCG conjugate. Atomic absorption spectrophotometry was used to measure the bound amount of silver and gold conjugates [65].

The basic principles of the lateral-flow technology continued to be refined through the early 1980s and were firmly established during the latter years of that decade, with the filing of several major patents on this technology format. Since then, at least another 500 patents have been filed on various aspects of the technology [63].

TABLE 3 **Detection Limits of Different Human Placental Lactogen Assays [65]**

		Detection Limit or Quantification	
Type of Label	**Method of Detection**	**pmol/L**	**fmol/well**
Sol (gold) particle	Naked eye	170	17.0
	Colorimeter	5.4	0.54
	Atomic absorption spectrophotometry	1.4	0.14
Enzyme	Naked eye	110	11
	Colorimeter	90	9
Radio isotope	Gamma counter	1.4	0.28

8. CONCLUSION

Immunology, histology, different clinical aspects, nanoparticle production, polymer synthesis and many other pure and applied developments contributed into evolution of rapid tests. This evolution mostly related with clinical assay; clinical assay adopts a lot of practical and fundamental achievements from clinical research. For example, labelling of antibodies through their conjugation with enzymes and metal nanoparticles were developed originally for histology.

One of the reasons why the methods based on antigen − antibody interaction become so widespread is that approach, once designed for a particular analyte, can be adapted with small modification to the analysis of various analytes. Changing of matrixes (eg, from urine to wheat extract) needs adjusting of sample pretreatment procedure, but neither the method itself. Table 4 presents some essential for rapid test development dates for 90 years till 1980 and introduction of lateral-flow immunoassay. The further progress is

TABLE 4 **Important Dates for Immunoassay Development**

Year	**Occurrence**	**Authors**
1890	Demonstration of antibody activity against diphtheria and tetanus toxins. Beginning of humoral theory of immunity	Emil von Behring and Shibasaburo Kitasato
1900	Antibody formation theory	Paul Ehrlich

Continued

TABLE 4 Important Dates for Immunoassay Development—cont'd

Year	Occurrence	Authors
1901	Complement fixation test	Jules Bordet and Octave Gengou
1906	Wassermann test for syphilis	August Paul von Wassermann
1938	Antigen–antibody-binding hypothesis	John Marrack
1939	Quantitative precipitation test Technique for antibody purification	Michael Heidelberger
1941	Haemagglutination assay and haemagglutination inhibition assay	George Hirst
1956	Latex agglutination test	Jacques Singer and Charles Plotz
1959	Radioimmunoassay	Solomon Berson and Rosalyn Yalow
1960	Immunological pregnancy test by haemagglutination inhibition	Leif wide and Carl Axel Gemzell
1966	Technique to fix antibody or antigen to the surface of a container	Wide and Jerker Porath
1966	Coupling antigens or antibodies with enzymes	Stratis Avrameas and José Uriel
1966	Coupling antigens or antibodies with enzymes	
1971	ELISA technique	Peter Perlmann and Eva Engvall
1971	EIA technique	Anton Schuurs and Bauke van Weemen
1972	Antibody specific to human chorionic gonadotropin. Radioimmunoassay for human chorionic gonadotropin	Judith Vaitukaitis
1978	Early pregnancy test (e.p.t.) commercially available for home application; it included a vial of purified water, an eye dropper, a test tube and an assortment of compounds including sheep's blood.	Warner-Chilcott
1979	Western blotting	George Stark Harry Towbin W. Neal Burnette
1980	Application of metal colloidal particles as a label for immunoassays	Jan Leuvering

related not only with the names of scientists, but also with breakthroughs in technology and the company's interests.

ACKNOWLEDGEMENT

The work was supported by Russian Science Foundation (project 14-13-00229).

REFERENCES

[1] S.S. Deshpande, Enzyme Immunoassays From Concept to Product Development, Chapman & Hall, London, 1996.
[2] J.H. Humphrey, R.G. White, Immunology for Students of Medicine, Blackwell Scientific, Oxford, 1964.
[3] M. Heidelberger, Chemical aspects of the precipitin and agglutinin reactions, Chem. Rev. 24 (1939) 323–343.
[4] M.E. Koivunen, R.L. Krogsrud, Principles of immunochemical techniques used in clinical laboratories, Labmedicine 37 (2006) 490–497.
[5] J.D. Howell, Technology in the Hospital: Transforming Patient Care in the Early Twentieth Century, Johns Hopkins University Press, Baltimore, 1995.
[6] H.P. Schwarz, F. Dorner, Karl Landsteiner and his major contributions to haematology, Br. J. Haematol. 121 (2003) 556–565.
[7] K. Landsteiner, Zur Kenntnis der antifermentativen, lytischen und agglutinierenden Wirkungen des Blutserums und der Lymphe Centralblatt f. Bakteriologie, Parasitenkunde u, Infektionskrankheiten 27 (1900) 357–362.
[8] J. Bordet, O. Gengou, Recherches sur le Coagulation du Sang et les Sérums Anticoagulants, Annales de l'Institut Pasteur 15 (1901) 129–145.
[9] P.M. Tiwari, Nobel Prize Winners of the World, Diamond Pocket Books (P) Ltd. X30, Okhla Industrial, 2013.
[10] J. Bordet, Les proprietes des sensibilisatrices et les theories chimiques de l'immunite, Ann. Inst. Pastuer, Paris 18 (1904) 593–632.
[11] A.M. Silverstain, History of Immunology, Academic Press, San Diego, 1989.
[12] G.K. Hirst, Adsorption of influenza hemagglutinins and virus by red blood cells, J. Exp. Med. 76 (1942) 195–209.
[13] R.W. Schlesinger, A. Granoff, George K. Hirst (1909–1994) Virology 200, 1994, p. 327.
[14] J.M. Singer, C.M. Plotz, The latex fixation test. I. Application to the serologic diagnosis of rheumatoid arthritis, Am. J. Med. 21 (1956) 888–892.
[15] F.J. Gella, J. Serraa, J. Generx, Latex agglutination procedures in immunodiagnosis, Pure Appl. Chem. 63 (1991) 1131–1134.
[16] L.B. Bangs, New developments in particle-based tests and immunoassays, J. Int. Fed. Clin. Chem. 2 (1990) 188–193.
[17] E. Harlow, D. Lane, Antibodies – a Laboratory Manual, Cold Spring Harbor Laboratory Press, Cold Spring Harbor, NY, 1988, p. 726.
[18] S.A. Berson, R.S. Yalow, Assay of plasma insulin in human subjects by immunological methods, Nature 184 (1959) 1648–1649.
[19] S.A. Berson, R.S. Yalow, A. Bauman, M.A. Rothschild, K. Newerly, Insulin-I131 metabolism in human subjects: demonstration of insulin binding globulin in the circulation of insulin treated subjects, J. Clin. Invest. 35 (1956) 170–190.

[20] S.A. Berson, R.S. Yalow, Species-specificity of human anti-beef, pork insulin serum, J. Clin. Invest. 38 (1959) 2017–2025.
[21] S.J. Goldsmith, Radioimmunoassay: review of basic principles, Semin. Nucl. Med. 5 (1975) 125–152.
[22] R.S. Yalow, S.A. Berson, Immunoassay of endogenous plasma insulin in man, J. Clin. Invest. 39 (1960) 1157–1175.
[23] S.A. Berson, R.S. Yalow, S.M. Glick, J. Roth, Immunoassay of protein and peptide hormones, Metabolism 13 (1964) 1135–1153.
[24] S.M. Glick, J. Roth, R.S. Yalow, S.A. Berson, Immunoassay of growth hormone in plasma, Nature 199 (1963) 784–787.
[25] S.M. Glick, J. Roth, R.S. Yalow, S.A. Berson, The regulation of growth hormone secretion, in: C. Pincus (Ed.), Recent Progress in Hormone Research, Academic Press, New York, 1965.
[26] D.S. Skelley, L.P. Brown, P.K. Beach, Review: radioimmunoassay, Clin. Chem. 19 (1973) 146–186.
[27] M.B. Jaffe, H.R. Behrman, Methods of Hormone Radioimmunoassay, Academic Press, New York, 1974.
[28] J. Landon, A.C. Moffat, The radioimmunoassay of drugs. A review, Analyst 101 (1976) 225–243.
[29] S.A. Berson, R.S. Yalow, General principles of radioimmunoassay, Clin. Chim. Acta 22 (1968) 51–69.
[30] A.H.B. Wu, A selected history and future of immunoassay development and applications in clinical chemistry, Clin. Chim. Acta 369 (2006) 119–124.
[31] L.E.M. Miles, C.N. Hales, Labelled antibodies and immunological assay systems, Nature 219 (1968) 186–189.
[32] K. Catt, H.D. Niall, G.W. Ttegear, Solid-phase radioimmunoassay of human growth hormone, Biochem. J. 100 (1966) 31c–33c.
[33] L. Wide, J. Porath, Radioimmunoassay of proteins with the use of Sephadex-coupled antibodies, Biochim. Biophys. Acta 30 (1966) 257–260.
[34] K.J. Catt, H.D. Niall, G.W. Tregear, A disc method for solid-phase radioimmunoassay of proteins, Aust. J. Exp. Biol. Med. Sci. 45 (1967) 702–706.
[35] K.J. Catt, G.W. Tregear, Solid-phase radioimmunoassay in antibody-coated tubes, Science 158 (1967) 1570–1572.
[36] R.H. Purcell, D.C. Wong, H.J. Alter, P.V. Holland, Microtiter solid-phase radioimmunoassay for hepatitis B antigen, Appl. Environ. Microbiol. 26 (1973) 478–484.
[37] M. Ceska, U. Lundkvist, A new and simple radioimmunoassay method for the determination of IgE, Immunochemistry 9 (1972) 1021–1030.
[38] R.M. Lequin, Enzyme immunoassay (EIA)/Enzyme-Linked immunosorbent assay (ELISA), Clin. Chem. 51 (2005) 2415–2418.
[39] E. Nieschlag, E.J. Wickings, A review of radioimmunoassay for steroids, Z. Klin. Chem. Klin. Biochem. 7 (1975) 261–271.
[40] S.G. Johansson, Radioimmunoassay of IgE and IgE antibody and its clinical application, J. Clin. Pathol. Suppl. Assoc. Clin. Pathol. 6 (1975) 33–37.
[41] L. Guilloux, S. Ricard-Blum, G. Ville, J. Motin, A new radioimmunoassay using a commercially available solid support for the detection of IgE antibodies against muscle relaxants, J. Allerg. Clin. Immunol. 90 (1992) 153–159. PMCID:PMC, pmid:1436078.
[42] D.R. Hoffman, Z.H. Haddad, Diagnosis of IgE-mediated reactions to food antigens by radioimmunoassay, J. Allerg. Clin. Immunol. 54 (1974) 165–173.

[43] T.A.E. Platts-Mills, P.W. Heymann, M.D. Chapman, M.L. Hayden, S.R. Wilkins, Cross-reacting and species-specific determinants on a major allergen from *Dermatophagoides pteronyssinus* and *D. farinae*: development of a radioimmunoassay for antigen P1 equivalent in house dust and dust mite extracts, J. Allerg. Clin. Immunol. 78 (1986) 398–407.
[44] R.C. Bast, T.L. Klug, E.S. John, E. Jenison, J.M. Niloff, H. Lazarus, R.S. Berkowitz, T. Leavitt, C.T. Griffiths, L. Parker, V.R. Zurawski, R.C. Knapp, A radioimmunoassay using a monoclonal antibody to monitor the course of epithelial ovarian cancer, N. Engl. J. Med. 309 (1983) 883–887.
[45] H. Kato, T. Torigoe, Radioimmunoassay for tumor antigen of human cervical squamous cell carcinoma, Cancer 4 (1977) 1621–1628.
[46] P. McBride, D. Bradley, M. Kaliner, Evaluation of a radioimmunoassay for histamine measurement in biologic fluids, J. Allerg. Clin. Immunol. 82 (1988) 638–646.
[47] E. Engvall, P. Perlmann, Enzyme-linked immunosorbent assay (ELISA). Quantitative assay of immunoglobulin G, Immunochemistry 8 (1971) 871–874.
[48] B.K. van Weemen, A.H.W.M. Schuurs, Immunoassay using antigen-enzyme conjugates, FEBS Lett. 15 (1971) 232–236.
[49] S. Avrameas, J. Uriel, Méthode de marquage d'antigènes et d'anticorps avec des enzymes et son application en immunodiffusion, C R Acad. Sci. Hebd. Seances Acad. Sci. D 262 (1966) 2543–2545.
[50] S. Avrameas, Coupling of enzymes to proteins with glutaraldehyde, Immunochemistry 6 (1969) 43–52.
[51] A. Voller, G. Huldt, C. Thors, E. Engvall, New serological test for malaria antibodies, Br. Med. J. 1 (1975) 659–661.
[52] I. Ljungstrom, E. Engvall, E.J. Ruitenberg, Proceedings: ELISA, enzyme-linked immunosorbent assay – a new technique for sero-diagnosis of trichinosis, Parasitology 69 (1974) xxiv.
[53] E. Engvall, Quantitative enzyme immunoassay (ELISA) in microbiology, Med. Biol. 55 (1977) 193–200.
[54] M. Seppala, E.M. Rutanen, M. Heikinheimo, H. Jalanko, E. Engvall, Detection of trophoblastic tumour activity by pregnancy-specific β1 glycoprotein, Int. J. Cancer 21 (1978) 265–267.
[55] B.K. van Weemen, The rise of EIA/ELISA, Clin. Chem. 51 (2005) 2226.
[56] A. Voller, D.E. Bidwell, A simple method for detecting antibodies to Rubella, Brit. J. Exp. Pathol. 56 (1975) 338–339.
[57] G. Wolters, L.P.C. Kuijpers, J. Kacaki, A.H.W.M. Schuurs, Enzyme-immunoassay for HbsAg, Lancet II (1976) 690.
[58] J. Renart, J. Reiser, G.R. Stark, Transfer of proteins from gels to diazobenzyloxymethyl-paper and detection with antisera: a method for studying antibody specificity and antigen structure, Proc. Natl. Acad. Sci. U.S.A. 76 (1979) 3116–3120.
[59] J.C. Alwine, D.J. Kemp, G. Stark, Method for detection of specific RNAs in agarose gels by transfer to diazobenzyloxymethyl-paper and hybridization with DNA probes, Proc. Natl. Acad. Sci. U.S.A. 74 (1977) 5350–5354.
[60] H. Towbin, T. Staehelin, J. Gordon, Electrophoretic transfer of proteins from polyacrylamide gels to nitrocellulose sheets: procedure and some applications, Proc. Natl. Acad. Sci. U.S.A. 76 (1979) 4350–4354.
[61] W.N. Burnette, "Western blotting": electrophoretic transfer of proteins from sodium dodecyl sulfate-polyacrylamide gels to unmodified nitrocellulose and radiographic detection with antibody and radioiodinated protein A, Anal. Biochem. 112 (1981) 195–203.

[62] E.M. Southern, Detection of specific sequences among DNA fragments separated by gel electrophoresis, J. Mol. Biol. 98 (1975) 503–517.
[63] B. O'Farrell, in: C. Wong, H.Y. Tse (Eds.), Chapter 1 Evolution in Lateral Flow–based Immunoassay Systems Lateral Flow Immunoassay Raphael, Springer, 2009.
[64] M. Horisberger, J. Rosset, H. Bauer, Colloidal gold granules as markers for cell surface receptors in the scanning electron microscope, Experientia 31 (1975) 1147–1149.
[65] J.H.W. Leuvering, P.J.H.M. Thal, M. van der Waart, A.H.W.M. Schuurs, Sol particle immunoassay (SPIA), J. Immunoassay 1 (1980) 77–91.

Chapter 3

Formats of Rapid Immunotests—Current-Day Formats, Perspectives, Pros and Cons

Chapter Outline

1. INTRODUCTION

There is a traditional continuum in rapid test technologies from highly accurate methods, requiring infrastructure and a centralized approach, to the technologies that can be used in a decentralized testing strategy that requires little to no supporting infrastructure. Because of big variety of tests, it is difficult to make clear classification of all current day and existed methods. Which property of the test could be the base of such classification? Nature of sample or analyte, type of label, immunoassay format, applied tools, automatization? Among big variety of possibilities, the classification of rapid test for homogeneous (no solid support at all) and heterogeneous (presence of any kind of solid support) looks like most fundamental and logical.

Comprehensive Analytical Chemistry, Vol. 72. http://dx.doi.org/10.1016/bs.coac.2016.03.012

Heterogeneous rapid tests could be classified on the base of solid support properties, without relation to the nature of the process, responsible for analytical signal appearance. Variety of solid supports could be divided as:

1. classical such as glass or plastic tubes, microtitre wells with procedure as shortened analogue of classical ELISA;
2. porous membranes and gels with attached immunoreagents and sample flowing through the pores;
3. micro- and nanoparticle colloidal systems.

The first group is also historically first to develop and now used in variants with reagents introduced inside tube/well (covalently bound or just dried) or reagent solution applied to the tube/well together with analyte solution.

The second group is the widest for the time. Principle of this group could be called in the terms of immunochromatography owing to combination of longitudinal or transverse liquid flow through a carrier with immunochemical reactions [1]. Fig. 1 (from Ref. [1]) illustrates the main kinds of immunochromatographic systems.

- The most widely used lateral flow test strips are designed to detect immune complexes in the binding areas (sector I-α).
- The immunoaffinity columns are used to concentrate analytes prior to detection in the binding areas (sector II-α).
- Lateral flow test strips are designed to detect washed-out compound (sector I-β).
- Immunoaffinity columns are designed to detect washed-out compound (sector II-β).
- Cross-flow membranes are designed to detect bound immune complexes (sector III-α).
- Cross-flow membranes are designed to detect washed-out compound (sector III-β).

The third group is situated close to genuine homogeneous methods and the progress in nano- and microcarrier development allows to reach big advance in the areas of automatization and improvement of sensitivity.

The second and third groups combine the speed of a homogeneous immunoassay with the separation of reacted and unreacted immunoreagents by using a solid supports.

Homogeneous rapid tests could be classified on the base of the nature of the process, responsible for analytical signal generation.

There are well-known approaches to improve characteristics of immunoassay, which are common for regular immunoassays and rapid tests:

- improvement of antibody affinity
- optimization of antibody cross-reactivity
- optimization of conjugates structure

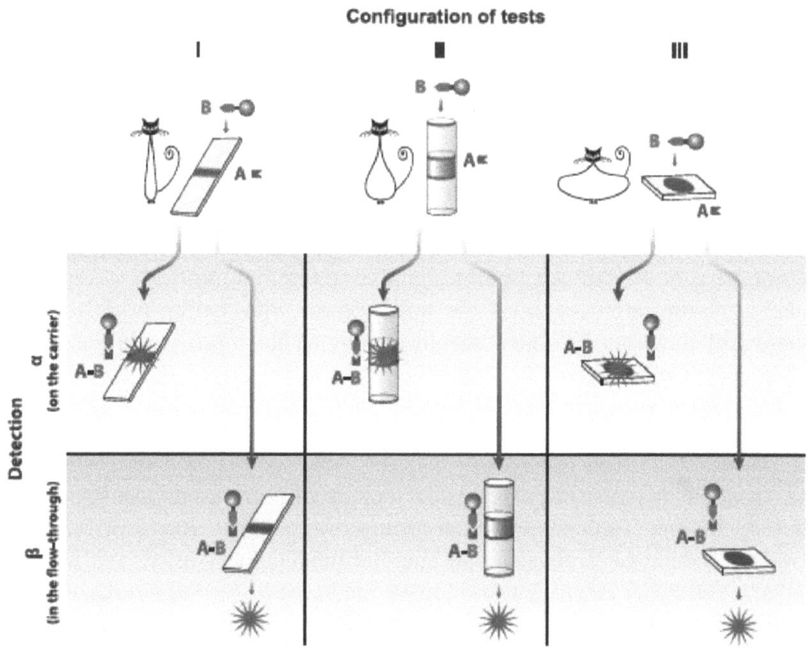

FIGURE 1 Test systems for immunochromatographic assays. A and B are interacting immunoreagents and their derivatives; compounds detected when recording the results of the assay are marked by red (dark grey in print versions) stars. Lateral flow test strips are designed to detect bound immune complexes (I-α); lateral flow test strips are designed to detect in the flow-through (I-β); immunoaffinity columns are designed to detect bound immune complexes (II- α); immunoaffinity columns are designed to detect in the flow-through (II- β); cross-flow membranes are designed to detect bound immune complexes (III- α); cross-flow membranes are designed to detect in the flow-through (III-β). *The image is from B.B. Dzantiev, N.A. Byzova, A.E. Urusov, A.V. Zherdev, Immunochromatographic methods in food analysis, Trends Anal. Chem. 55 (2014) 81–93.*

- maintaining of immunoreagents stability
- optimization of label and method of signal registration special for rapid tests

A peculiarity of rapid tests is a short assay time, so kinetic of all processes (separation, migration, binding, signal generation) should be fast enough.

It is important to mention that the properties of antibody could be exposed differently depending on the assay format. The most proteins tend to adopt a conformation in which hydrophilic groups tend to be on the outside of the protein and hydrophobic groups within it, it seems inevitable that hydrophobic binding to polymer surfaces causes conformational changes in the adsorbed protein. Antigens coated onto a solid phase may lose critical epitopes, either by conformational changes, because they are sterically hidden, or because the epitope is predominantly hydrophobic. New epitopes may also arise that were previously hidden. Antibodies also undergo conformational changes, affecting not only the number of active binding sites, but also their affinity for antigen. It is by no means uncommon to find an antibody that works well in solution but not when bound to solid phase [2].

2. HETEROGENEOUS METHODS

Each type of heterogeneous immunoassay has depended on a separation of unbound tracer before the bound signal is measured. Without a separation (such as a thorough wash of the solid phase with buffer prior to signal generation) the level of signal would always be the same, regardless of the concentration of analyte.

2.1 Rapid ELISA Tests

The speeded up versions of classical enzyme-linked immunosorbent assay (ELISA) are one of the most popular groups of rapid tests. Rapid ELISA test is a popular format of 'wet-lab' type analytic biochemistry assay and has been used as a relatively fast diagnostic tool in medicine as well as a quality control test in various industries. As for other types of immunoassay, performing an ELISA involves at least one antibody with specificity for a particular antigen. An antigen could be of different nature; target antibody of interest also could play the role of an antigen and so could be detected by immunoassay techniques. Although chemical coupling of immunoreagents to solid phases is possible, many solid-phase immunoassays rely upon noncovalent hydrophobic adsorption. Horseradish peroxidase (HRP) and alkaline phosphatase are the most popular enzymes for labelling. Fluorescent labels and other alternatives to enzyme-based detection can be used for plate-based assays too. Despite not involving enzymes as reporters, these methods are also generally referred to as a type of ELISA. Besides, wherever detectable probes and specific protein-binding interactions can be used in a plate-based method, these assays are often called ELISAs despite not involving specific antibodies. The ELISA formats vary for low-molecular-weight analytes (competitive formats) and high-molecular weight analytes (sandwich format) [2–4]. Classification of rapid ELISA test formats is presented on Fig. 2.

For ***low molecular weight analytes***, ELISA is realized in competitive format. This is common when the antigen is small and has only one antibody-binding site (epitope). Variations of this method consist of labelling of the antigen and labelling of the antibody.

Formats with labelled antibody: The conjugate of antigen with coating protein adheres to the bottom of well on microtitre plate, or glass, or plastic tube. The mixture of sample to analysis and specific antibody is applied to the well. An antigen in solution competes with an attached to the solid surface antigen for binding with restricted amount of primary specific antibody in

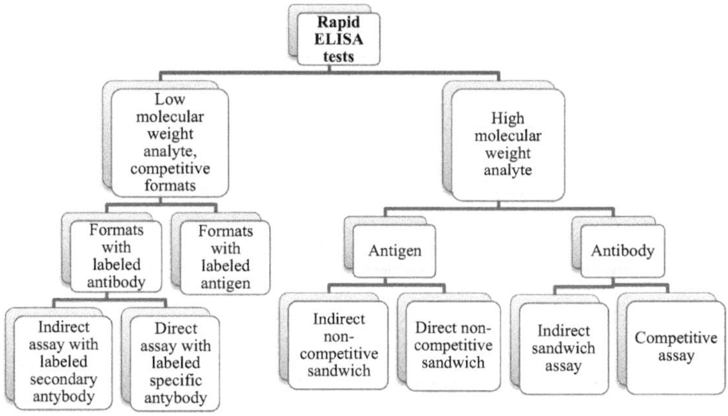

FIGURE 2 The scheme of rapid ELISA test principles.

solution. In the absence of antigen, in the sample, a primary specific antibody binds to the target attached to the solid surface, and a labelled secondary antibody binds to the primary for detection. Intensity of the colour, developed after chromogenic substrate application, is inversely proportional to the amount of analyte in the sample (Fig. 3). Because of detected parameter is an amount of attached to the solid surface labelled secondary antibody, this format often terms indirect competitive ELISA. The main advantage of an indirect ELISA with labelled secondary antibody is that one, labelled secondary antibody can be used with many different primary antibodies. The disadvantage which is especially important in the terms of rapid tests development is a requirement of this antibody application, which means of additional steps, including incubation and washing step.

Labelling with enzyme of primary specific antibody allows to avoid these additional steps, but requires labelling procedure during method development. Also, immunoreactivity of the primary antibody might be adversely affected by labelling with enzymes or tags. Intensity of the developed colour is inversely proportional to the amount of analyte in the sample and indicates amount of attached to the solid surface amount of labelled primary specific antibody; this format is called direct competitive ELISA (Fig. 4). Direct detection is not widely used in classical ELISA but is quite common for immunohistochemical staining of tissues and cells.

Format with labelled antigen: The specific primary antibody adheres to the bottom of well on microtitre plate, or glass, or plastic tube. The mixture of sample to analysis and labelled with enzyme antigen is applied to the well. An antigen in solution competes with labelled antigen for binding with restricted amount of primary specific antibody attached to the solid surface. In the presence of antigen in the sample, it binds to the specific antibody (Fig. 5, left). Otherwise, in the absence of antigen in the sample, a labelled antigen binds to

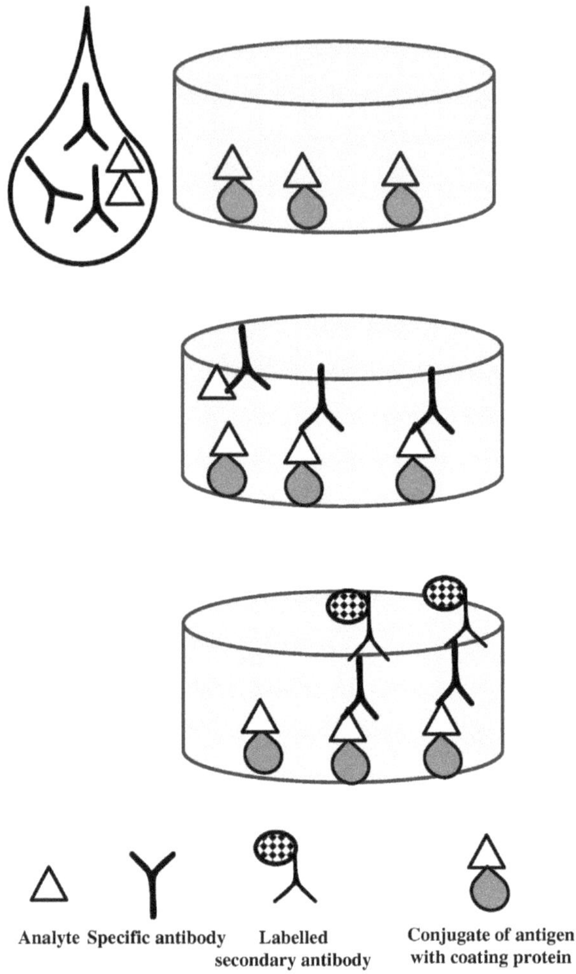

FIGURE 3 Competitive indirect ELISA.

the specific antibody (Fig. 5, right). A decrease in signal indicates the presence of the antigen in samples when compared to assay wells with labelled antigen alone. As for above-mentioned formats, an intensity of the colour, developed after chromogenic substrate application, is inversely proportional to the amount of analyte in the sample.

For ***high-molecular-weight analytes***, ELISA is more effective in noncompetitive sandwich format. These assays benefit from being 'reagent excess' systems and offer precise and fast assays equally applicable to complex automated immunoassay machines through to dipsticks.

Sandwich ELISAs typically require the use of matched antibody pairs, where each antibody is specific for a different, nonoverlapping part (epitope)

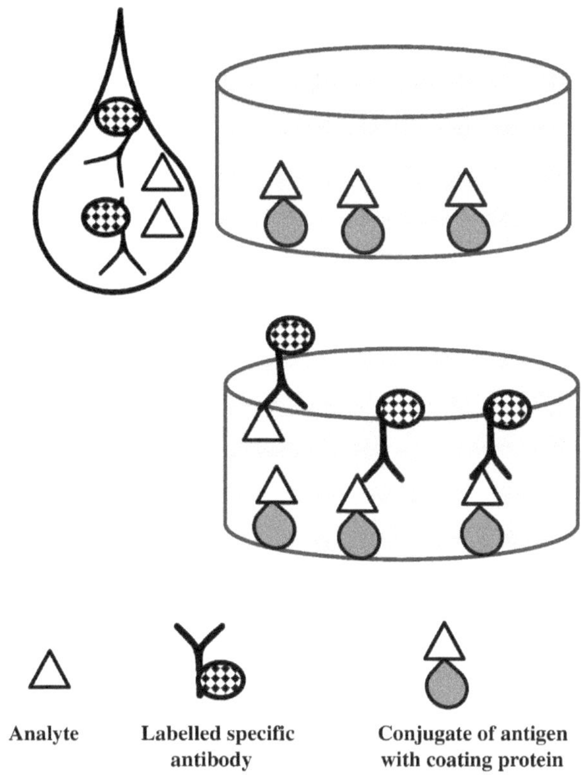

FIGURE 4 Competitive direct ELISA.

of the antigen molecule. The first antibody, termed the capture antibody, is coated to the solid support. Next, the sample solution is added to the wells and analyte, if present, attached to surface. Addition of a second antibody layer, termed the detection antibody, follows this step in order to measure the concentration of the analyte (Fig. 6). If the detection antibody is labelled with an enzyme, then the assay is called a direct sandwich ELISA. If the detection antibody is unlabelled, then a labelled secondary antibody will be needed resulting in an indirect sandwich ELISA.

In some cases, an immunoassay may use an antigen to detect for the antibodies present; the analyte may be an antibody rather than an antigen. For detection of receptors, including antibody, while they are high molecular weight, special ELISA formats exist.

Sandwich indirect format: The sample antibody is sandwiched between the antigen coated on the plate and a labelled secondary antibody (Fig. 7). Intensity of the analytical signal is directly proportional to the amount of bound sample antibody. The more antibodies present in the sample, the stronger the

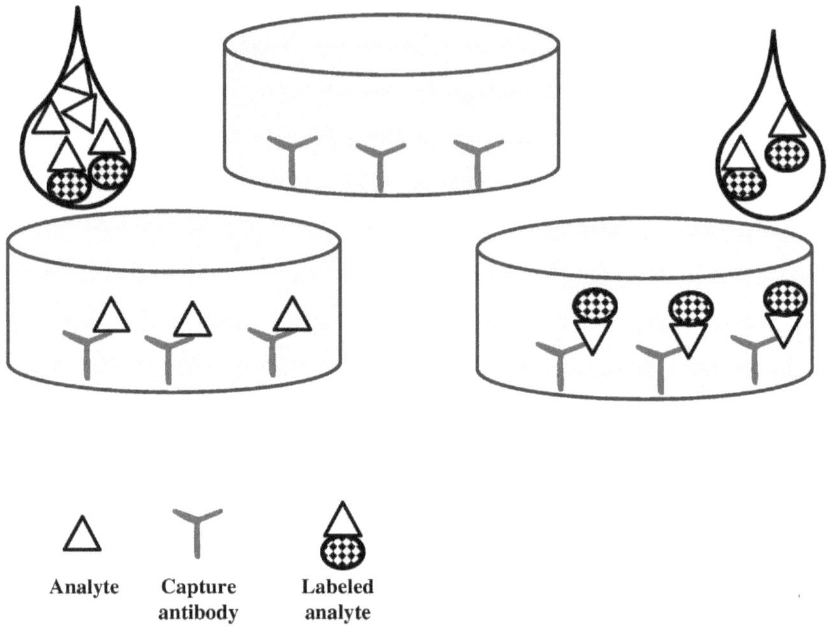

FIGURE 5 Competitive ELISA with labelled analyte. Left: presence of analyte; right: absence of analyte.

colour development. This format is suitable for determining total antibody level in samples.

Competitive format: The specific sample antibodies compete with the labelled antibodies. Intensity of the colour is inversely proportional to the amount of bound sample antibody. The more antibodies present in the sample, the less labelled antibody bound and less analytical signal would be observed (Fig. 8).

With high automatization and high throughput for many laboratories, ELISA is very convenient method to perform analysis, but for some cases duration of assay is extremely important. There are several approaches to reduce ELISA duration and to obtain results faster.

The first and the most obvious one is to reduce duration of incubation steps. Standard ELISA protocol includes several incubation steps with duration often requiring more than 1 h each, for coating reagent overnight incubation often used. Immunoassays involve a binding reaction between an analyte and at least one antibody. This reaction takes from a few seconds to many hours to achieve equilibrium, depending on a range of different factors. Early immunoassays often had overnight incubations to allow the reaction to fully reach equilibrium. This makes ELISA not suitable to be termed 'rapid test' and not applicable, for example, for urgent analysis in clinical assay or for quality

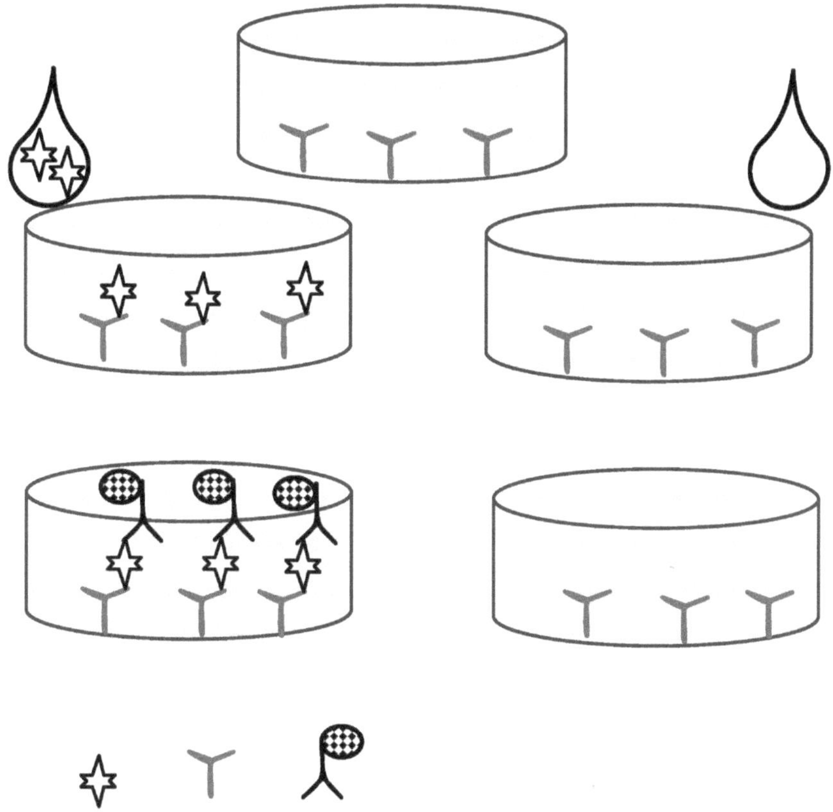

FIGURE 6 Noncompetitive sandwich ELISA. Left: presence of analyte; right: absence of analyte.

control of perishable commodities. The most obvious way is shortening of the incubation period. Most current immunoassays involve comparatively short incubations and do not allow the reaction to reach a state of equilibrium. The longer time for incubation, the closer results of the process to the equilibrium conditions and the less deviation for obtained results could be expected. Obviously, decreasing of the incubation time could result in lowering of reproducibility even in the case of shaking the plates and 37°C incubation.

The design of the assay influences the equilibration time in a number of ways. For example, if the antibody is immobilized on the surface of microtiter plate wells, the average distance that an analyte molecule has to move before coming into contact with a molecule of antibody is much longer, comparing with homogeneous assays or assays utilizing micro- and nanoparticles as solid support. The pH, ionic strength, and temperature also affect the reaction time

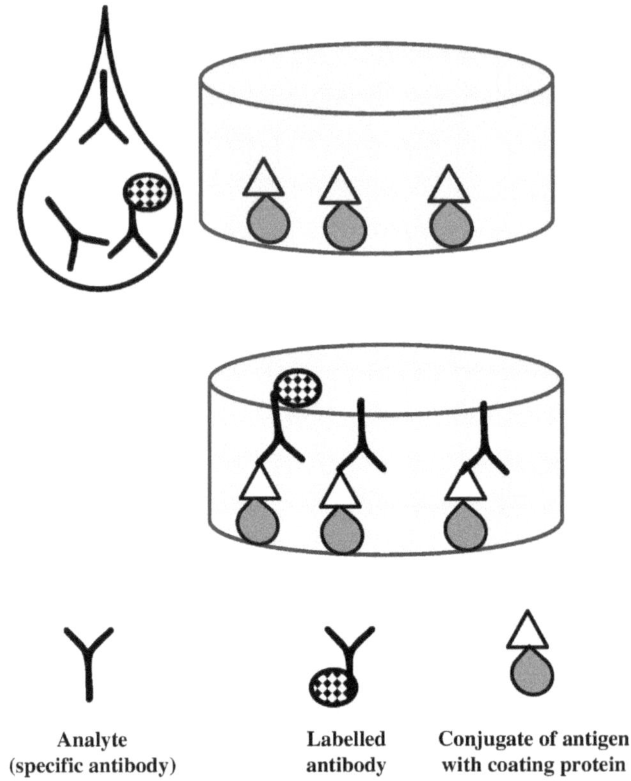

FIGURE 7 Competitive sandwich ELISA for antibody detection.

[2]. According to some reports, external influence, such as microwave irradiation is able to reduce ELISA incubation time [5]. For commercial products, duration of incubation times should be followed exactly as stated in the product manual to ensure optimal ELISA test performance.

Another way of shortening of the total procedure time is to reduce amount of steps. For example, preincubation of covering reagent into well or membrane-backed wells allows to reduce incubation and washing step. This kind of tests with precoated plates is available commercially for big variety of analytes and samples. Direct ELISA instead of indirect ELISA safe time for labelled secondary antibody application and followed washing step.

Progress in biotechnology and nanoscience provides possibility to use alternative ways of signal generation and alternative labels. While enzyme-based ELISA is extremely sensitive, especially in chemiluminescent variant, enzyme application is related with incubation with chromogenic, fluorescent or chemiluminescent substrate to perform enzyme-catalysed process, which is time-dependent and sensitive to pH, temperature and presence of additional compounds. Different alternative tags were developed: gold, silver, selenium,

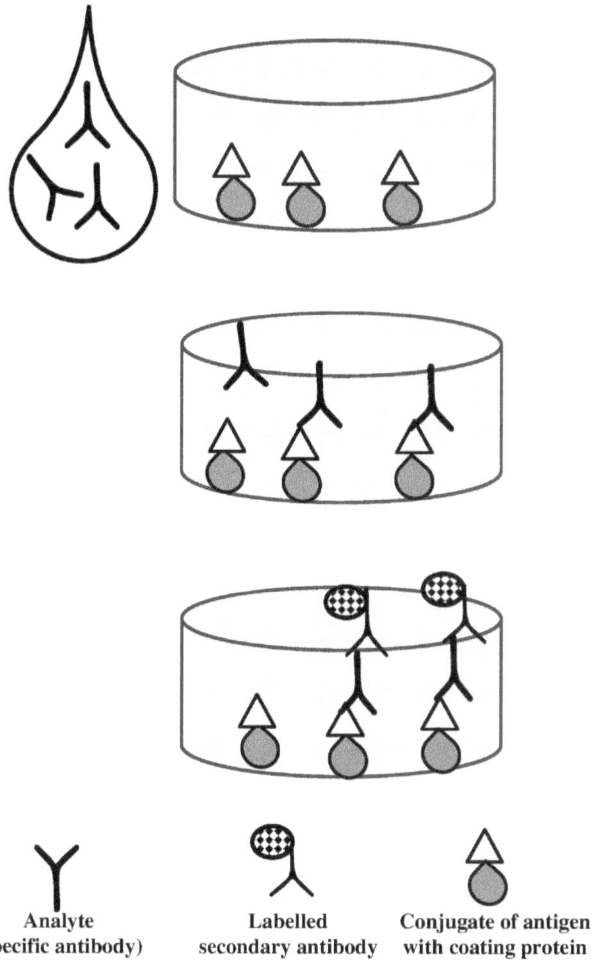

FIGURE 8 Noncompetitive sandwich ELISA for antibody detection.

carbon (both highly dispersed and structured such as carbon nanotubes) and silica particles, latex, dyes nanoparticles, Raman reporters, conductive metal particles, luminescent molecules and complexes, quantum dots (QDs), lanthanide-based long-living phosphors, up-converting phosphors and liposomes loaded with different agents.

2.2 Membrane-Based Tests

In the last decade, there has been a continuous growth in the development of rapid methods for mycotoxin analysis. Moreover, noninstrumental rapid screening techniques that could be used outside the laboratory environment, at the place of sampling, are becoming increasingly more important.

Simplest result evaluation of membrane-based tests is visual. Therefore, different visual labels are used, such as enzymes for catalytic enzymatic reactions, colloidal gold, fluorescent labels and liposomes, encapsulating a visible or fluorescent dye. The basic immunoassay formats used in noninstrumental tests are competitive or noncompetitive ELISAs. Visual detection usually give qualitative results as positive/negative (yes/no), characterizing the presence (or absence) of the target analyte in concentrations higher than the fixed cut-off level. The cut-off level can be established based on either noticeably reducing of colour development or complete colour suppression. Some tests assume semiquantitative estimation based on the colour intensity. For competitive immunoassays there is a problem to distinguish of the response at low analyte concentration from the response at zero analyte concentration, as both conditions give rise to large signals. Also, by definition, competitive assays are not 'reagent excess' in nature. Limited concentrations of both capture antibody and competing partner must be employed, with the resulting limitations for assay speed. In addition, not only must reagent concentrations be limited, but they must also be precisely maintained across an assay series to achieve acceptable results. Even relatively modest variations of concentration of the competitive reagents can have a profound effect on the precision of the competitive assay. Considering the very wide range of small molecule analytes where fast, high-performance assay systems are required in, for example, human and veterinary medical, forensic and defence, and environmental and quality assurance applications within the water, food, and beverage industries, these are critical problems that required a solution [2]. To make interpretation of results easier, special control zones have been included in the majority of present-day tests [6].

Membrane-based tests unite advantages of homogeneous and heterogeneous methods. It combines the speed of a homogeneous immunoassay with the separation of reacted and unreacted compounds by a variety of heterogeneous methods. Another advantage is that the fluid flow through the carrier membrane and separates of reacted from unreacted products without the need for additional precipitation or washing steps. All membrane-based tests are combined with a simple sample preparation procedure: in the case of clinical samples it is dilution of samples or its direct application on the tests; in the case of food samples it is usually extraction with methanol or methanol/water (or buffer) mixture, filtering and dilution with buffer.

2.2.1 Lateral Flow Immunoassay

Lateral flow immunoassays (LFIA) attract attention not only as one of successful rapid tests, but also as state-of-the-art of user friendly test for both home application and clinical laboratory. Combination of bio-science, nanotechnology and technology of material construct basis for tests used in human and veterinary medicine, food and beverage manufacturing, pharmaceutical, medical biologics and personal care product manufacturing, environmental remediation, and water utilities. Lateral flow tests are well suited to replace laboratory-based immunoassays in decentralized testing locations, first of all for point-of-care purposes.

LFIA or immunochromatographic strips are widely used for the early detection of pregnancy, for drug screening, to identify markers for various diseases, and for a number of other analytical tasks. Several very informative reviews have been recently published [1,7]. Few decades ago, the term 'immunochromatography' was used to describe not the kind of test methods we mean in modern days. It was setup and type of analysis based on separation of samples on a column containing a sorbent with covalently bound antibodies specific to a target analyte(s) [8,9]. This approach called immunoaffinity columns and is used in modern analytical practice to separate and to concentrate various substances usually before chromatography-based analysis [10].

Today, the most widespread immunochromatographic system is the test strip—an assembly of several plain porous carriers impregnated with immunoreagents [1]. A typical LFIA format consists of a surface layer to carry the sample from the sample application pad via the conjugate release pad along the strip encountering the detection zone up to the absorbent pad [7].

Essential in the current LFIA, the movement of a liquid sample, or its extract containing the analyte of interest, along a strip of polymeric material thereby passing various zones where molecules have been attached that exert more or less specific interactions with the analyte and detectable immune complexes formation in certain zones of the test strip [7,11].

The basic principle of LFIA as a combination of chromatography (separation of components of a sample based on differences in their movement through a sorbent) and immunochemical reactions emerged in early 1980s and were firmly established during the latter years of that decade. Since then it has been implemented in many different ways. The main advantage of LFIA is toughly user friendly format—absence of any hand steps and manipulation with reagents. In practical realization of assays, all necessary reagents are preimpregnated onto strip compounds. The fluid flow through the carrier enables separation of reacted from unreacted products and sample compound without the need for additional precipitation or washing steps.

Immunoreagents' immobilization on a carrier and fluid flowing through that carrier allow for:

- adjustable and rapid formation of immune complexes;
- removal of unreacted compounds from the binding zone during the analysis and
- the use of special zones to concentrate and detect target complexes [1].

LFIA Process. LFIA are prefabricated strips of a carrier material containing dry reagents that are activated by applying the fluid sample. When a test is run, sample is added to the proximal end of the strip and to a sample application pad. Here, the sample is treated to make it compatible with the rest of the test. Alternatively, the labelled analyte or recognition element(s) are dried in a reaction tube and the sample plus strip are added to the tube [7]. The treated sample migrates through this region to the conjugate pad. Labelled analyte or recognition element(s) (depending on the assay format) are dried on this pad and after addition of the sample, this material will be remobilized by the fluid flow; specific interactions will be initiated here and will continue as both migrate into a porous membrane (Fig. 9).

On the membrane within test and control zones the other specific biological components of the assay have been immobilized. These are typically proteins, either antibody or antigen, which have been laid down in bands in specific areas of the membrane where they serve to capture the analyte and conjugate as they migrate by the capture zones (lines). Excess reagents move past the capture zones and are entrapped in the wick or absorbent pad. Results are interpreted on the reaction matrix as the presence or absence of lines of captured conjugate, read either by eye or using a reader [7,12].

Test construction. The LFIA, once the poor relation to laboratory-based tests, is being viewed more and more as a truly versatile technology, capable of more than adequate performance at all ends of the diagnostic continuum. In order for this evolution in capability to occur, there has been a continuous improvement in materials, reagents, approaches to development, manufacturing equipment, manufacturing process technology, and the introduction of a whole new generation of facilitative technologies [12].

The carrier membranes are often thin and fragile, so they are attached to a plastic or nylon basic layer to simplify cutting and handling. The membranes

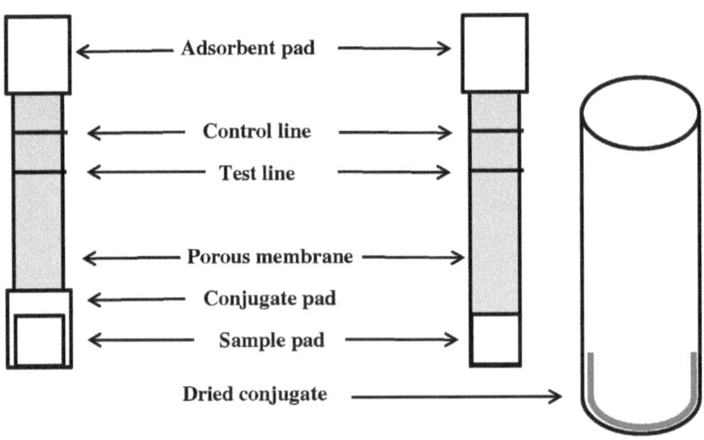

FIGURE 9 Scheme of lateral flow immunoassays strip. Left: dry chemistry variant; right: labelled reagent in separate tube.

are produced from nitrocellulose, nylon, polyethersulfone, polyethylene or fused silica. At one end of the strip a sample application pad is provided. For same assay this pad is also able to perform sample pretreatment, for example, separation of red blood cells from the sample. The sample application pad is usually made of cellulose or cross-linked silica.

The liquid moves because of the capillary force of the strip material, but to maintain a flow an absorbent pad is attached at the distal side of the strip. This absorbent pad will wick the liquid to the end of the strip, thus maintaining the flow [7]. For some commercial tests, especially for in-home patient application, robustness is achieved by housing the strips in a plastic holder, where only the sample application window and a reading window are exposed.

At least two lines are sprayed on the strip: a test line and a control line. At the test line the recognition of the sample analyte and the reporter will result in the required response [7]. A response at the control line confirms a proper flow of the liquid through the strip and complex formation. Usually appearance of control line is a sufficient evidence of the efficacy of the test. More than one test lines can be applied for follow goals.

Quantitative (semiquantitative) evaluation of the response also allow for multianalyte testing [13]. Application of several test lines with the same immunoreagents provides possibilities for better evaluation of the analyte concentration, as it was shown for analytes of clinical [14] and food safety [15] interests. Different cut-off levels for certain analytes could be needed, for example, because of different legislation limits in different countries. So, there is necessity to adjust sensitivity of tests according to new discoveries in clinical, food, environmental assay and to changes in the maximum permissible levels of certain contaminants (in food assay). Variation of immunoreagents concentration allowed to vary cut-off levels of the assays. For example, changing of the dilution of analyte-covered protein conjugate on the test line and specific antibody concentration resulted in changing of the analyte (deoxynivalenol) indicator ranges from 250–500 to 1000–2000 μg/kg [16]. If concentration of immunoreagent on the test lines is the same, the number of lines appearing on the strip is directly proportional to the analyte concentration, as it was shown for prostate acid phosphatase sandwich format determination [17] and aflatoxin B1 with competitive format [15], urinary human serum albumin [14].

Simultaneous detection of several analytes. System for multianalyte testing can detect two or more types of antigens (and/or antibodies) in a single assay device at the same time. During the development of method, it is possible to establish for each analyte required sensitivity and type of reaction. A multianalyte lateral flow technique was developed for the simultaneous detection of two micotoxins (deoxynivalenol and zearalenone). The technique gave accurate and reproducible 'Yes–No' results with cut-off levels of 1500 and 100 μg/kg for deoxynivalenol and zearalenone in grain samples, respectively [18]. Similarly, the test with the visual detection limits for ochratoxin A and zearalenone with cut-off levels at 2.5 and 5 ng/mL for ochratoxin A and zearalenone, respectively was described by Shim et al. [19]; the visual detection of pesticides carbaryl and endosulfan with detection limits at 10 and 1 μg/kg, respectively was described by Zhang et al. [20]. Comparison of two sets of immunoreagents for detection of two analytes (pesticides carbofuran and triazophos) showed that the application of gold-labelled bispecific monoclonal antibody against both carbofuran and triazophos showed less sensitivity than the anti-carbofuran monoclonal antibody and anti-triazophos monoclonal antibody separately labelled with colloidal gold [21]. LFIA with two test lines allows to detect different forms of one compound (free and total prostate specific antigen in serum) [22], but from the technical point of view here was just LFIA for two different analytes, because for each form, own set of immunoreagents was used.

LFIA examples. Semiquantitative detection of an early cardiac marker, heart-type fatty acid–binding protein and an established risk marker for heart attack, C-reactive protein were realized in 'digital-style' lateral flow assay to confirm or exclude suspected acute myocardial infraction patients and simultaneously to identify apparent nonacute myocardial infarction patients at risk of developing cardiovascular events at an early stage [23].

The oligonucleotide lateral flow immunoassay device is composed of colloidal gold-labelled antibodies and oligonucleotide-labelled antibodies fixed in a conjugate pad, and the complementary oligonucleotide-labelled proteins are immobilized on a nitrocellulose membrane. If the target antigen is present in a specimen, the colloidal gold-labelled antibody and oligonucleotide-labelled antibody make a complex with the antigen. Subsequently, the formed complex migrates to the place where complementary oligonucleotide is immobilized and is bound to the solid phase via the DNA–DNA interaction. As a result, more than two types of reactions can be detected on a single assay device by the combination of colloidal gold-labelled antibodies, different oligonucleotide-labelled antibodies and complementary oligonucleotide-labelled proteins immobilized at different places on a nitrocellulose membrane [24].

Labels for LFIA. Enzymes were used as label for LFIA historically and still in use nowadays [25] Enzyme immunochromatography as a form of quantitative immunoassay was pioneered by Zuk et al. [26]. Later, enzyme immunoassay was realized in conventional LFIA format with specific reagents immobilized in a form of narrow bands [27]. In enzyme LFIA, the most often used label is HRP [25,28–30]; however, alkaline phosphatase [27,31] and cholinesterase [32] were also employed. As far as HRP is concerned, colourimetric [28–30], chemiluminescent [33,34] and electrochemical [35] detection modes were reported. Substrate solution contained 3,3′,5,5'-tetramethylbenzidine (TMB), and was predominantly employed for colourimetric detection in HRP-based LFIA [29]; however, 4-chloro-1-naphthol was also used in few studies [26,30].

But the most popular labels are nanoparticle based [7,36]. Labels are made of coloured or fluorescent nanoparticles with sizes of 15–800 nm, allowing an unobstructed flow through the membrane. They are most often made of colloidal gold [37–40] or coloured latex [41], less often selenium [42], carbon [43] or liposomes [44,45] are used. In liposomes, coloured [41], fluorescent [46] or bioluminescent [45] dyes can be incorporated, allowing visualisation, and, when applicable, quantitation of the response. The newest labels include QDs [47] and up-converting phosphor reporters [7,48,49]. For more information about labels for rapid tests, including LFIA, see "Chapter 4, Labels for optical immunotests" by Goryacheva [130].

Formats of LFIA. Competitive (or competitive inhibition) and noncompetitive (also known as sandwich or d**ouble antibody sandwich** or immunometric or direct) formats have been described for LFIA. These formats are determined by the analyte structure. For analytes with more than one epitope, the noncompetitive format with two different antibody is applicable. For small analytes, only competitive formats are possible.

Sandwich format of LFIA. This format is used when testing for larger analytes with multiple antigenic sites, such as human chorionic gonadotropin, Dengue antibody or antigen, or human immunodeficiency virus. The analyte should have two antigenic sites that can be recognized simultaneously by the two antibodies. Therefore, such assays cannot be used for low molecular weight analytes such as simple steroids, small peptides, most drugs, pesticides, mycotoxins, persistent organic pollutants.

In this format the test line is prepared using one analyte-specific antibody (Fig. 10). The conjugate release pad contains a second, labelled, anti-analyte antibody. Analyte in the sample will bind during the initial chromatographic process to the second antibody. The free epitope can bind to the immobilised antibody at the test line [7]. The response is directly proportional to the amount of analyte in the sample; the calibration curve (in the case of quantitative detection) will have linear view.

After liquid sample dissolves labelled specific antibody on the conjugate pad, analyte in the sample and these antibody will form immunocomplex which bind during the initial chromatographic process to the specific antibody on the test line by means of free antigenic sites on the surface of analyte molecule. Moving with the sample flow immunocomplex will concentrate on the test zone. This process is 'visible' (literally or figurative, depending on the label kind). If no analyte in the sample, no immunocomplex will be formed. Free-labelled antibody and/or part of immunocomplex-labelled antibody–analyte move further to the control zone, there they will bind and concentrate by secondary antibody. Thus, test line will be visible only in the presence of analyte in the sample, control line, otherwise should be visible at any case.

To obtain better sensitivity noncompetitive assay, utilize all reagents in excess. Unlike competitive assays, they are not largely dependent on antibody

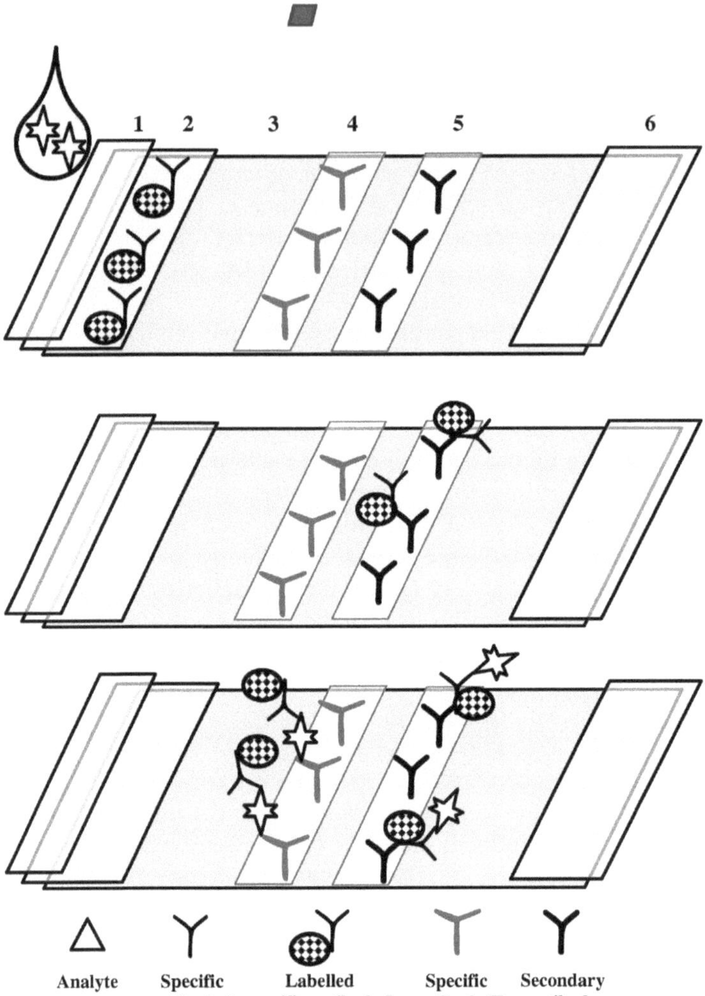

FIGURE 10 Noncompetitive sandwich lateral flow immunoassays. Top: sample application. Middle: absence of analyte — only control line is coloured. Bottom: presence of analyte — test and control lines are coloured. 1 -sample pad; 2 —conjugate pad; 3 — porous carrier membrane; 4 — test line; 5 — control line; 6 — adsorbent pad.

affinity. If problems such as high nonspecific binding of the label, degradation of assay specificity, and interference from heterophilic antibodies (all from the use of excess reagents) are minimized, then very low detection limits can be achieved by maximizing the signal-to-noise ratio of the label [50]. The two-site sandwich assays, in which the antigen is sandwiched between two specific antibodies and their combined selectivity determines the specificity, are inherently more specific than the single-site assays [50]. Less than an excess of sample analyte is desired, so that some of the labelled antibody–analyte complexes will not be captured at the test line and will continue to flow toward the control line.

Competitive formats of LFIA. When the analyte is of low molecular weight and has only one epitope, the format is restricted to the competitive design. Two variants are possible:

1. Antibody is sprayed at the test line, a mixture of sample analyte and labelled analyte is applied at the conjugate pad and the sample analyte and labelled analyte compete for binding sites on the antibody at the test line [51]. After liquid sample dissolves labelled analyte on the conjugate pad, they will flow with liquid stream to test line and control line. If analyte present in the sample, it will bind to the specific antibody on the test line. If no analyte in the sample, labelled analyte will bound to the test line and make it 'visible'. In both cases the rest of labelled analyte will move further to the control line and bind there. Thus, test line will be visible only in the absence of analyte in the sample, control line, otherwise should be visible at any case (Fig. 11).
2. An analyte–protein conjugate is sprayed at the test line, and a mixture of labelled antibody and sample analyte is applied at the conjugate pad, giving the sample analyte a head start for binding to the antibody [7,52]. As in sandwich format, after liquid sample dissolves labelled specific antibody on the conjugate pad, analyte (if present) and the antibody will form immunocomplex which will move with the front of liquid along membrane strip. Test lane contends bound analyte (usually analyte–protein conjugate), which will bind free labelled antibody. In analyte present in the sample, all labelled antibodies are bound with it and will not form immunocomplex on the test line. If amount of antibody is less than the amount of analyte, all antibodies epitopes will be occupied and no colour on the test line will appear. If amount of antibody is higher than amount of analyte or no analyte in the sample, free-labelled antibody will form immunocomplex with the analyte–protein conjugate sprayed at the test line and will 'visualize' test line. Free-labelled antibody and/or part of immunocomplex-labelled antibody analyte move further to the control zone, there they will bind and concentrate by secondary antibody. Thus, the test line will be visible only in the absence of analyte in the sample, the control line, otherwise, should be visible at any case (Fig. 12).

The preferred layout of competitive LFIA is dependent on the particular application. In the competitive LFIA formats the analytical signal (intensity of signal on the test line) is negatively correlated to the analyte concentration (ie, more analyte present, less signal; no analyte gives the highest signal). The calibration curve is sigmoidal shaped [7]. For realization of competitive assay the amount of labelled reagent on the conjugate pad should be chosen very carefully. Excess of the labelled reagent will result in necessity of high amount

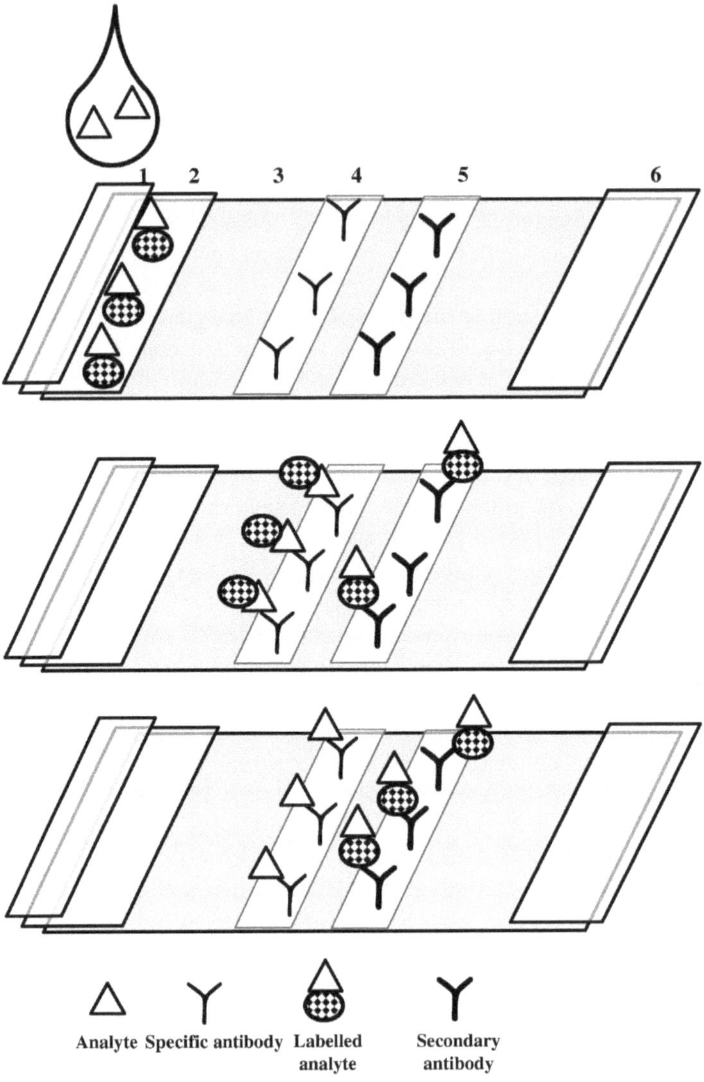

FIGURE 11 Competitive lateral flow immunoassays with labelled analyte. Top: sample application. Middle: absence of analyte – test and control lines are coloured. Bottom: presence of analyte – only control line is coloured. 1–6 – see Fig. 10.

of analyte to bind in immunocomplex and so sensitivity would be decreased. Shortage of the labelled reagent will result in a very weak analytical signal from the test line and also weak signal from the control line, which will complicate results interpretation.

LFIA analytical characteristics improvement. To improve LFIA characteristics along with the standard for all immunoassay approaches, such as

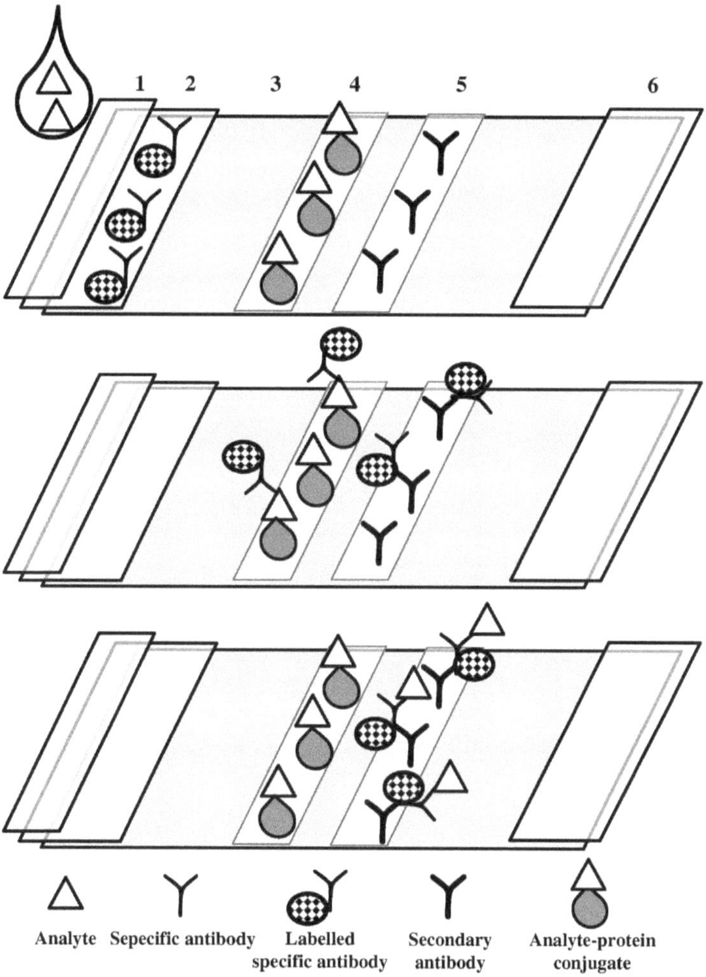

FIGURE 12 Competitive lateral flow immunoassays with labelled antibody. Top: sample application. Middle: absence of analyte – test and control lines are coloured. Bottom: presence of analyte – only control line is coloured. 1–6 – see Fig. 10.

improvement of antibody affinity and the immutoreagents stability, optimization of antibody cross-reactivity, conjugates structure, label properties and the method of signal registration, the important role playing the size and surface properties of the label and geometry of the membrane pores.

The sensitivity and the specificity of LFIA are controlled by the choice of immunoreagents. Currently, there are many different antibody preparations for almost all important antigens that are available from several manufacturers. The general recommendation is to choose antibodies with the highest affinity. However, to express immunochromatography methods, the priority is not the equilibrium but the kinetic binding constant. The antibodies also differ in their resistance to denaturing agents, which must be considered when working with organic extracts for food samples analysis, for example, refer Ref. [1].

Next, it is important to optimize the proportions of reagents to obtain a satisfactory low limit of detection (LOD) and produce reliable assay results. However, it is difficult to achieve a low LOD and high accuracy simultaneously, so some compromises must be made. For sandwich analyses, a higher concentration of immunoreagents will increase the number of complexes that can be detected at low concentrations of the antigen. However, this may promote nonspecific binding of the label. For competitive analyses in which there is an inverse relationship between the concentration of the analyte and the detectable label, low concentrations of immunoreagents will give a lower LOD [53]; however, this is accompanied by low signal intensity and decreased accuracy [1].

The time of the interaction between the reactants in homogeneous reactions (in the moving front of liquid) and in heterogeneous reactions (crossing binding zones in test kits) is another important factor. The time must be long enough for the immunochemical reactions to be complete. The duration of the interaction can be adjusted by selecting a suitable carrier and reaction medium [1,54]. The use of fine-pore nitrocellulose membranes and lower concentrations of detergents can enhance specific interactions, thus giving a lower LOD. However, fine-pore membranes are unsuitable for analyses of high-molecular-weight antigens, such as viruses and microbial cells [1].

Areas of application. Immunochromatography assays are rapid and simple, allowing for point-of-care testing. These advantages explain their success in medical diagnostics. Immunochromatography test strips are mass produced, and are widely used to detect a big variety of analytes with clinical interests, such as cardiac markers and infectious microorganisms. They are also used in serodiagnostic analyses to identify antibodies against various pathogens [1]. LFIA are POC immunoassays for self-testing that are sold over-the-counter and those performed by health-care professionals in hospital laboratories, hospital wards, clinics, community health centres and physician offices. Besides human medicine, LFIA satisfy demands of veterinary medicine, food and beverage manufacturing, pharmaceutical, medical biologics and personal care product manufacturing, environmental remediation, and water utilities. Lateral flow tests are also available and in development for biowarfare agents and pathogens such as anthrax, smallpox, avian influenza and other potential biological weapons [12]. The availability of established technologies to synthesize reagents and the relatively inexpensive equipment required to produce test strips (dispensers, cutters, laminators) mean that it is relatively easy to produce immunochromatographic kits for different applications. Details about preparation detailed in Refs [1,7,12].

Quantitative LFIA. Initially, immunochromatographic tests were designed for qualitative analysis, to provide information about whether the concentration of a certain analyte exceeded a certain limit. The noninstrumental estimation of results is based on visual evaluation. Visual tests usually give qualitative results as positive/negative (Yes/No), characterizing presence (or absence) of the target analyte in concentrations higher than the fixed cut-off level. This cut-off level could be established on the basis of either noticeably reducing of colour development or complete colour suppression. Presence of control zone makes interpretation of results easier.

However, at present, there is greater demand for quantitative or at least semiquantitative immunoassays. Some tests assume semiquantitative estimation on the basis of the colour intensity. For quantitative detection, the amount of the bound label is determined by digital image processing or by measurements based on other properties, such as fluorescence, electroconductivity or magnetism, instead of a simple presence/absence visual test result [1]. Even if qualitative 'Yes/No' information is all that is required, detectors provide documentation of results and eliminate subjective interpretations. Despite the increased costs of analysis, many manufacturers are therefore now producing immunochromatographic test strips combined with portable detectors: Roche Diagnostics (Switzerland), LRE Medical (Germany), Detekt Biomedical (United States), Qiagen (Germany), Alverix (USA) and Axxin Inc. (Australia). Photometric detectors are dominant in the market at present. Alere (USA) has developed a portable fluorescence lateral flow reader. The transition from qualitative to quantitative LFIA resulted in 3- to 50-fold improvements in the LOD [1].

2.2.2 Flow-Through Immunoassay

Flow-through or immunofiltration assay (IFA) or enzyme-linked immunofiltration assay (ELIFA), or dot-immunoassay utilizes regular ELISA formats. To perform flow-through (or vertical flow) assay the immunoreagents are attached to the membrane as separate spots or zones. Nonspotted surface of the membrane serves as an indicator of the presence or absence of nonspecific interaction of immunoreagents and sample components with the membrane surface. The membrane is placed on an absorbent body (filter paper, cellulose or cotton pad), which inhibits the immediate back-flow of fluids that could obscure results. The sample is added to the upper surface of the membrane where it flows through the membrane into the pad of absorbent material. Contrary to LFIA, nanoparticles [55,56] or colloidal dyes [57] are not commonly used as a label for flow-through assays. Enzymes are the most popular labels; even they require a chromogenic substrate application [58–60].

For low molecular weight analytes, detection competitive formats are used.

If specific antibodies (test zone) are attached to the membrane, an analyte (if presents) binds to the antibody spot on the membrane. After the washing step with small amount of buffer, the analyte–label conjugate is added and bound by the remaining unbound antibody (Fig. 13). The final step is the

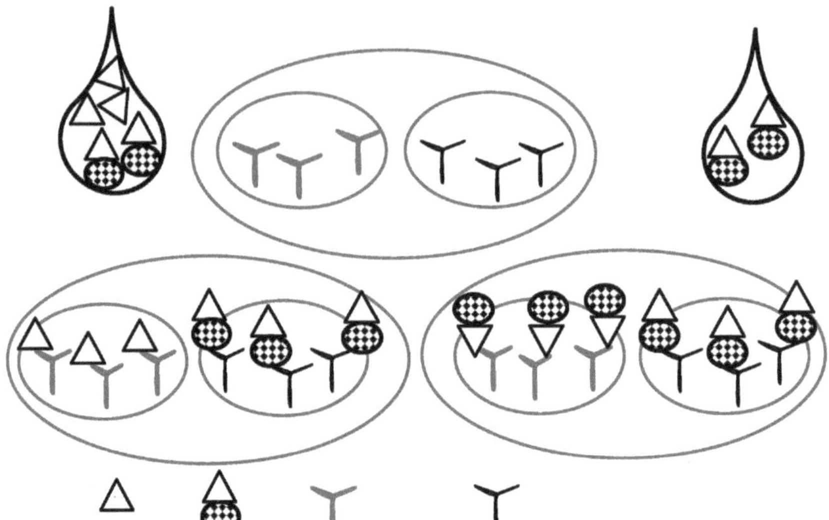

FIGURE 13 Flow-through membrane competitive immunoassay with labelled analyte. Test zone with specific to analyte antibody and control zone with specific to label antibody are shown. Left: presence of analyte – only control zone is coloured; right: absence of analyte – test and control zones are coloured.

chromogenic substrate application to obtain coloured product in the presence of enzyme [6], forestalled with second washing step [58–60]. As the control zone anti-enzyme antibody spot (in the case of enzyme label) is used.

If conjugated with covering protein analyte is attached to the membrane (test zone), cocktail of preincubated labelled specific antibody with the sample is added to the membrane surface. In the case of presence of analyte, it binds with the specific antibody-binding sites and the antibody will not remain on the membrane surface after washing step. In the case of analyte absence, labelled specific antibody will bind to the test zone. As the control zone the spot of secondary anti-species antibody is used [56]. The spot colour can be visually evaluated or instrumentally measured [61] and for competitive assay it is inversely proportional to the analyte concentration. As for LFIA, results could be considered as positive on the base of colour absence or colour partially suppression.

For high-molecular-weight analytes, noncompetitive formats are used. For antibody detection in schistosomiasis, immunoassay consisted of soluble egg antigen of *Schistosoma japonicum* coated onto membrane, mounted on a flow-through test device to provide the assay capture matrix. Soluble egg antigen absorbed to a red colloidal dye served as the antigen–antibody complex detecting reagents [57].

Comparison of detection sensitivity with the same set of immunoreagents showed lower carbaril LOD for flow-through immunoassay, comparing with LFIA (50 and 100 μg/L, respectively) [56]. Schneider et al. [62] compared dipstick and flow-through formats for fumonisin B1 and concluded that both tests showed the same sensitivity in buffer solutions (7.5−10 ng/mL). These tests, however, were 50 times less sensitive than the corresponding ELISA with instrumental detection. For sporidesmin A determination, it was also shown that a direct competitive ELISA was more sensitive. The lower sensitivity of a competitive immunoassay using visual evaluation is mainly the result of the high antibody density at the membrane dot required for sufficient colour development for a negative sample [63]. For brucella-specific antibodies, it showed a similar level of sensitivity of IFA to ELISA and it was more sensitive than both the rose bengal test and the complement fixation test for bovine brucellosis diagnosis [64,65].

To increase the sensitivity of flow-through assays, some modifications were applied. In particular, to avoid limitations related to volumes of washing buffer, reagents and sample 'open' construction of a membrane-based test was proposed, in which the absorbent body was not fixed to the reaction membrane. This modification makes it possible to wash with a stream of washing buffer. Adsorbent body replacement allowed applying several portions of sample resulting in preconcentration of target analyte in the antibody spot. No limitation in volume allowed to use signal amplification with biotinylated tyramine and avidin−HRP conjugate for aflatoxin B1 determination. This procedure decreased the limit of detection from 5 to 0.25 pg/spot (0.2−0.01 ng/mL) with densitometric detection [66]. For the simultaneous screening of several samples, they used four membrane strips with four separate antibody-coated test zones for each. A similar test with 36 spots was used for membrane-based flow-through immunoassay for T-2 toxin detection [6,67].

To eliminate matrix interferences a clean-up step was included in the assay. This allowed using less diluted sample extracts, which resulted in an improvement of the detection limit [68]. To improve the quality of immunoreagent spots on the membrane, a new spotting method was developed by Saha et al. [69]. The flow-through immunoassay has been developed for simultaneous estimation of aflatoxin B1 and ochratoxin A in chilli samples [70].

To make assay really multiplex, a prototype eight-well immunofiltration test device was developed for the simultaneous determination of seven mycotoxins: aflatoxin B1, fumonisin B1, T-2 toxin, roridin A, deoxynivalenol, diacetoxyscirpenol and ochratoxin A. Membrane sections coated with different anti-toxin antibodies were fixed to each well. Under the membranes a filter pad was attached. The assay procedure was similar to one for a single toxin determination, except that a mixture containing the respective toxin−enzyme conjugates was used for all wells [71]. As for other types of multianalyte tests with visual detection, there was a loss of sensitivity in comparison with single analyte determination, particularly because the demands in assay simplification and easy interpretation did not allow the use of optimal reagent concentrations [6].

A simple methanol-based extraction followed by filtration and dilution steps was prescribed [59]. The tolerance of 40% methanol was found to be due to the application of small-size (0.8 mm diameter) spots on membranes, as the tolerance decreases to 20% with gradual increase in spot size [70]. For some matrices, however, flow-through assays need additional clean-up steps before analysis, such as an immunoaffinity clean-up for aflatoxin M1 detection in milk [60], or a solid-phase clean-up for ochratoxin A detection in roasted coffee [72].

2.3 Column Tests

The replacement of a carrier membrane with the 3D solid support, such as specially prepared gel or porous polymer carrier, with bound immunoreagents makes it possible to enhance the sensitivity of tests. Analysis is performed in a transparent plastic column (as for solid-phase extraction), and it includes the stages of consecutively passing the sample, immunoreagents and washing buffers through the bulk of the carrier, placed on the bottom of the column. An incorporation of the immunoreagents into the bulk of the carrier makes it possible to concentrate analytes using the principle of immunoaffinity columns. The column immunoassay format combines efficient analyte separation, preconcentration, and sensitive detection in a small test cartridge using stepwise incubations [73,74].

For low-molecular-weight analytes, detection competitive format with labelled analyte is used (Fig. 14). The specific primary antibody attaches to the solid support directly or through immunocomplex with secondary anti-species antibody. Then the sample to analysis passed through the column and analyte, if present, concentrates on the antibody-binding sites. Applying next labelled antigen binds with remained nonoccupied antibody-binding sites. An intensity of the colour, developed after chromogenic substrate application, is inversely proportional to the amount of analyte in the sample [73,74]. Results were obtained as detection immunolayer colour development (for negative result) or no colour development (for positive result) after performing a direct competitive immunoassay. For high-molecular-weight analytes, noncompetitive sandwich format is used [75]. To create affinity carrier covalent binding of antibody to CNBr-activated Sepharose [73,74,76] or physical adsorption on the polymer [75,76] was used.

Integration of solid-phase clean-up and immunoassay in one device was obtained by the clean-up tandem immunoassay column, which contained a clean-up layer and detection immunolayer(s). As a clean-up layer, aminopropyl-derived silica was used [73]. The purpose of the clean-up layer was to reduce the extract colour intensity and to minimize matrix effects. The clean-up layer could be positioned above or below the detection immunolayer, or

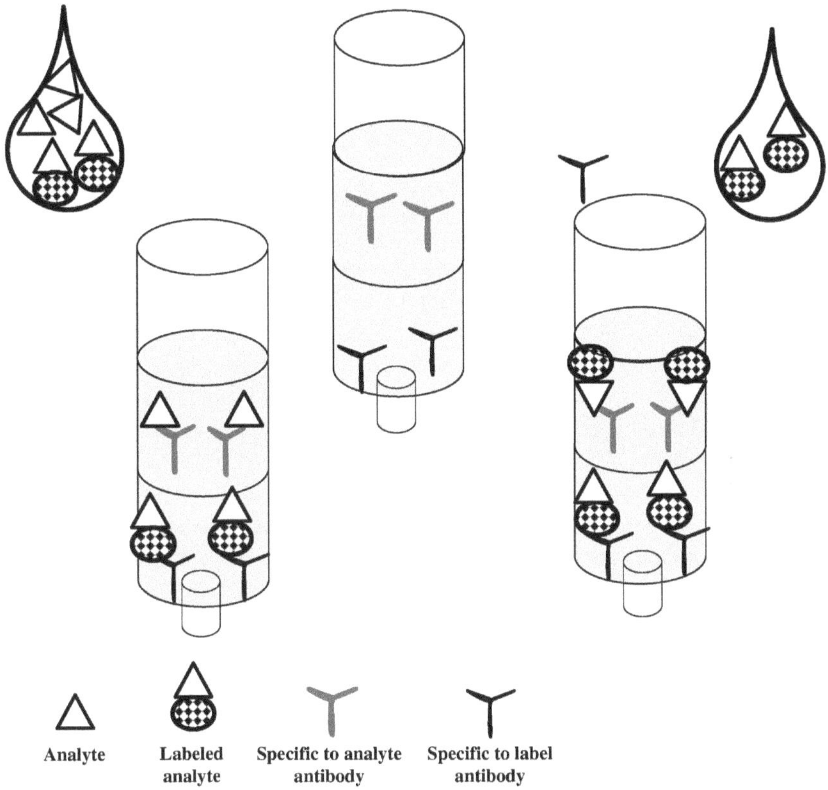

FIGURE 14 Flow-through column competitive immunoassay with labelled analyte. Test layer with specific to analyte antibody and control layer with specific to label antibody are shown. Left: presence of analyte – only control layer is coloured; right: absence of analyte – test and control layers are coloured.

inside separate disposable after sample application column [6]. To simplify the analytical procedure, a conjugate could be introduced into the column. The analysis procedure itself could consist of seven [77,78], five [73] or three steps [79], including washing steps.

For multiple analytes screening the number of detection immunolayers was increased. For each detection immunolayer a specific antibody for the separate analyte was bound to the gel before placing it into the column. This approach was realized for the simultaneous detection of two mycotoxins in *Capsicum spp.* spices: nutmeg, ginger, black pepper and white pepper [80]. As for membrane-based multiple immunoassays with visual detection, to prevent incorrect interpretation of results, time and intensity of developed colour should be about the same for both analytes [6].

Assay sensitivity could be varied by changing antibody and conjugate dilutions, and also volume of extract or liquid sample. Column design had no limitation for volumes of liquid sample or extract, reagent and washing buffers. It allowed, in particular, to use high extract volumes to improve assay sensitivity. In particular, to obtain a control level of 0.2 μg/L for ochratotin A in beer, a 12 mL of beer were used, whereas 10 ml of skimmed milk was used to reach a control level of 0.04 μg/L in the determination of aflatoxin M1 in milk [81].

For quantitative results, evaluation photometer reader was developed by Senova GmbH (Germany). As for other kinds of tests quantitative assay enabled to decrease LOD [76]. Mostly enzyme labelling (HRP) was used for signal detection for analysis of different analytes like mycotoxins, trinitrotoluene and polycyclic aromatic hydrocarbons [82] in various samples such as natural and drinking water [74,83], wine [84], spices [80], milk [85] and feed [86]. Beloglazova et al. [82] compared three different kinds of labels, ie, HRP, colloidal gold and luminescent QDs with respect to rapid visual on-site testing of benzo[a]pyrene in drinking water. The assays based on the use of particle labels require four consecutive working steps only, while those based on HRP require five additions of reagent. The lower limit of detection for benzo[a]pyrene is 5 ng/L in case of using enzyme or QDs as a label, but 25 ng/L when using gold nanoparticles.

3. HOMOGENEOUS IMMUNOASSAY

Homogeneous immunoassays do not need any kind of solid support and require only mixing of a sample and immunochemical reagents followed by detection. Immunochemical binding produces a physically detectable analytical signal that obviates the need to separate bound from free label. This is the principal difference with heterogeneous methods, there a separation of unbound reagents before measurement of the signal of bound to the solid surface complex is an essential phase of assay. Without a separation (washing step or flow with liquid stream) the label amount would always be the same, regardless of the analyte concentration. To make a separation not required, the label (or pair of two labels) should generate the signal which is sensitive to the immunocomplex formation: binding to the analyte in a noncompetitive assay or to the antibody in a competitive assay. In other words, the unique and common characteristic of each of these methods is that they provide a mechanism for modifying the signal produced by a label as a function of an immunochemical binding event. So theoretically all nanoscale assemblies are capable of continuously monitoring concentrations of target species in a sample through changing of any kind of analytical signal as a response to immunocomplex formation could be reporter system. Labels play role of active nanosensor, sensitive to immunocomplex formation. This contrasts with heterogeneous methods, which depend on evaluation of signal intensity, which is directly related with an amount of the label. This makes labels and their properties the most suitable base for homogeneous immunoassay classification. In this section, only properties of labels related to homogeneous assay are described. Detailed discussion of labels is presented in Chapter "Labels for optical immunotests" by Goryacheva [130]. Homogeneous methods have been developed for high- and low-molecular-weight analytes using both competitive and noncompetitive assays.

Homogeneous immunoassays could be classified as assays without labels and label-based assays. The first reported immunoassays were homogeneous but did not employ a label. They are attributed to Kraus, who in 1897 coined the term precipitin for the precipitate formed upon mixing an antigen and an antibody [87]. Likewise, the earliest immunoassays that employed a label were also homogeneous [88,89].

Because the rate of the binding reaction is not limited by slow diffusion to a surface, incubation times are short, usually only a few seconds to a few minutes; and the nonseparation assay protocols minimize the requirements for automation. Immunochemical binding can be followed either kinetically or after achievement of binding equilibrium. Because the separation and washing steps of heterogeneous methods are essentially more error prone and tend to influence weak binding reactions with an attendant decrease in sensitivity, theoretically homogeneous methods are more sensitive than heterogeneous immunoassays. Practically, as sample constituents are not removed by a wash step, variations in signal caused by nonspecific effects of the sample matrix prevent realization of this potential advantage [89].

3.1 Latex Agglutination and Metal Nanoparticles Agglutination

Typically antibodies (or antigen) are bound to the surface of the particles, which form aggregates when a polyvalent antigen (or target antibody) is present. Both competitive and noncompetitive sandwich assays can be used. A noncompetitive sandwich assay can be set up with two antibody-coated particles that bind a multivalent antigen (Fig. 15). The more antigen that is present, the more agglutination occurs. Latex agglutination results take about 15 min to an hour.

Alternatively, in competitive assay added antigen can inhibit antibody-induced aggregation of antigen-coated particles. This 'agglutination inhibition' method is generally less sensitive. In each case, it is only necessary to combine the sample and reagents and optically measure the agglutination [89]. The agglutination and agglutination inhibition schemes are presented on Fig. 16. The latex agglutination test is a wet-lab method to check for certain antibodies or antigens in a variety of bodily fluids including saliva, urine, cerebrospinal fluid or blood. Numerous techniques have been described for agglutination tests. The tests could be performing using slides, test tubes or microtitre plates.

The slide test is used primarily for screening large amount of sera. An undiluted or single-diluted serum sample is used for this test. In this case, results would be qualitative. For semiquantitative results, serum should be

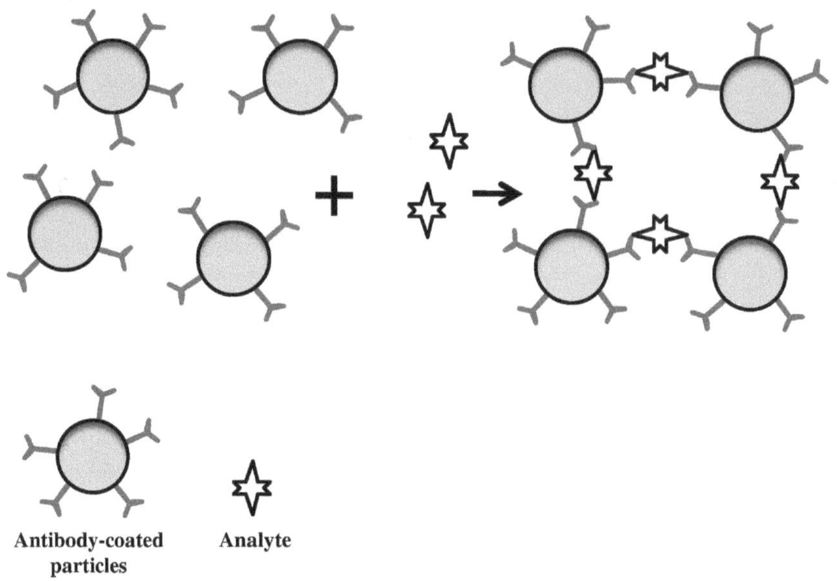

FIGURE 15 Noncompetitive sandwich latex immunoagglutination assay: antibody-coated particles agglutinated by antigen molecules.

serially diluted. Slide tests provide results in a matter of minutes. Latex particles for this test could be coloured to enhance readability. For performing of the tests, the reagents are mixed on the glass slides or disposable test cards. An employment of white disposable test cards with coloured latex particles facilitates detection of agglutination. One of the problem with using the slide test is related with application of nondiluted samples: this is potential for false negatives. If target antibody concentration in undiluted sample is too high, the reaction with a constant amount of antigen results in prozone phenomenon, and a false negative result. Therefore, serial dilutions should be done for samples from patients whose clinical symptoms are highly indicative even if a screening test of particular disorder is negative. To reduce the probability of a false negative result, many tests use a single 1:20 dilution [90]. Serum samples that are positive in the slide test could be analysed semiquantitatively in mictotitre plates. In this assay a constant amount of antigen is added to the serially diluted sample in the wells of microtitre plate. Endpoint titres are reported as the reciprocal of the highest dilution of sample in which clamping occurs. Comparing with tubes, microtitre plates facilitate operations with several dilutions. As well, a minimal amount of reagents is required for each well [90].

Visual observation of latex agglutination is moderately sensitive and not quantitative. Most instrumental quantitative measurements rely on turbidimetry or nephelometry. Turbidimetry measures the intensity of a beam of light

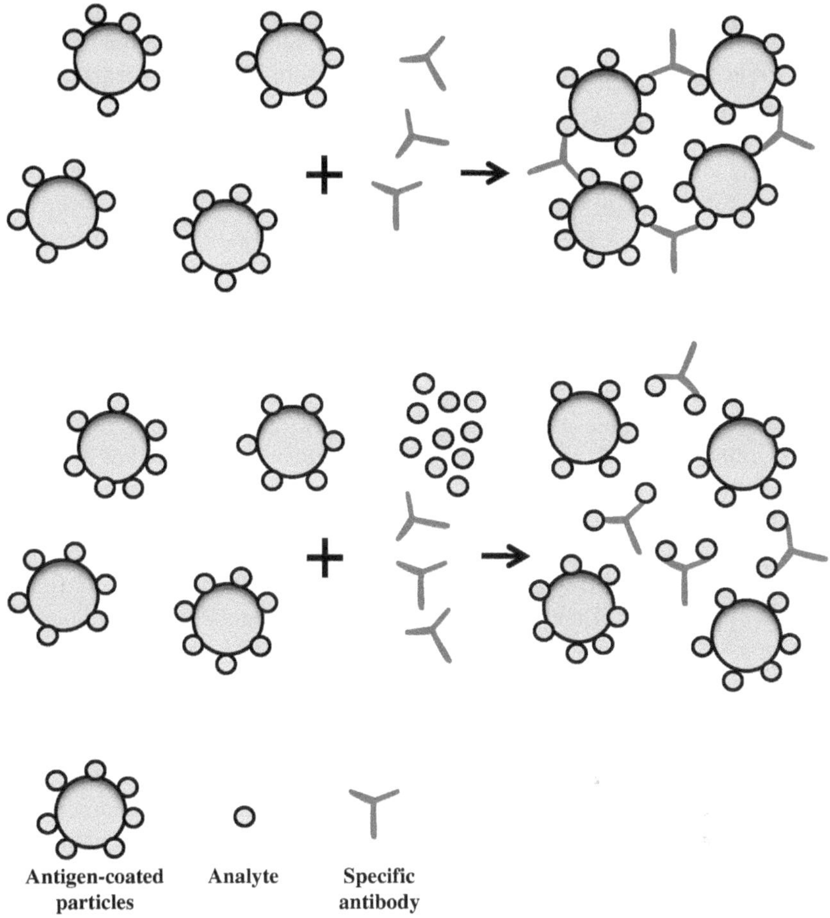

FIGURE 16 Agglutination (for antibody detection) and agglutination inhibition (for competitive antigen detection) schemes.

transmitted through the sample. Nephelometry measures the light that is scattered at an angle away from the beam. Nephelometry is more sensitive but is more subject to interference from particulate matter in the sample [89]. Comparison with ELISA on the example of human immunodeficiency virus detection showed better sensitivity for latex agglutination in non-HIV patients [91]. In the other study, ELISA did not appear to have any advantage over latex agglutination test for diagnosis of rheumatoid arthritis [Banchuin1992] and cryptococcal antigen [92].

High sensitivity can be achieved by measuring the angular anisotropy of laser light scattered from a suspension of particles (laser nephelometry). The light scattering is a sensitive function of the particle size, particularly when the wavelength of light is on the order of the size of the particles. The method provides greater sensitivity than simple nephelometry, but interference from particulate material in the sample and reagents is also increased [93].

New approaches to develop techniques for latex agglutination measurement include electrochemical impedance spectroscopy. A new class of impedance-based lab-on-chip immunosensors in which the immunoagglutination of latex in the presence of target antigens leads to precipitation from solution between interdigitated microsized electrodes on a chip [94].

Many commercially available tests use latex particles coated with immunoreagents of interest. The agglutination tests can be used to determine blood type, detect anti-self antibodies and antibodies to variety of infectious agents. Among commercially available tests are tests for fibrin(ogen) degradation products and D-dimer assays for diagnosis of disseminated intravascular coagulation, detection of *Cryptococcus neoformans*, diagnosis of bacterial meningitis, leptospirosis, rheumatoid arthritis, visceral leishmaniasis, to identify *Candida dubliniensis* colonies and *Staphylococcus aureus*.

Metal nanoparticles have high indices of refraction and produce particularly strong light scattering relative to latex. The scattering intensity is strongly dependent on nature of metal and particle size. The special term solid particle immunoassay (SPIA) is related to these methods. SPIA is a term generally applied to those assays in which the colloidal particles are inorganic; however, organic particle may also be solid. The most popular material for nanoparticle is gold. Additionally with other advantages the important one for immunoagglutination property is changing of the colloid colour. Then gold nanoparticles are isolated from each other, each nanoparticle appeared pink-red colour as a result of surface plasmon resonance. Then as result of agglutination, as the distance become smaller the colour of colloid changes for grey-blue and further clumping leads to colour disappearance. This colour change could be easily detected by naked eye. Gold nanoparticle concentrations in the low femtomolar range can be detected.

3.2 Fluorescence Polarization Immunoassay

Fluorescence polarization immunoassay (FPIA) is a homogeneous competitive immunoassay based on the increase in fluorescence polarization (FP) of fluorescent-labelled small antigens when bound by specific antibody. Walter Dandliker first described the application of FP for detection of antigen–antibody interactions in the 1960s [95,96]. Apposite to the agglutination methods, FPIA is mostly the method for detecting low-molecular-weight analytes. The principle of detection is competition of analyte and labelled with fluorescent label analyte (tracer) for restricted amount of antibody-binding sites in solution (Fig 17). Tracers for FPIA are generally contained as label small organic dye fluorescein.

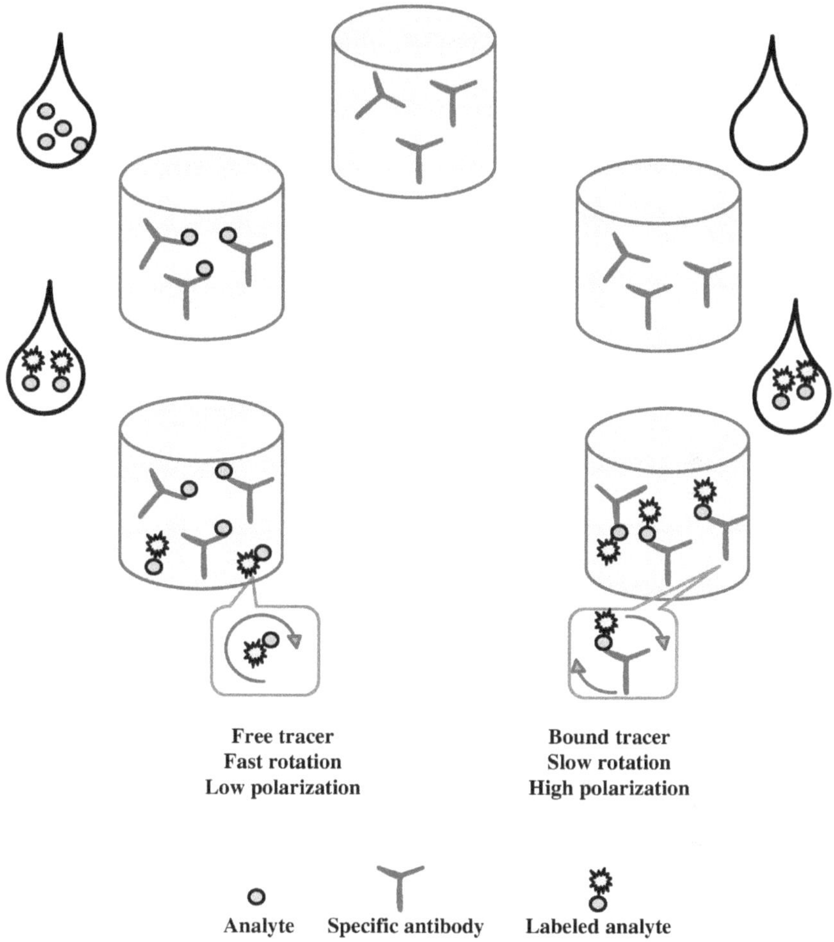

FIGURE 17 Fluorescent polarization immunoassay. Left: presence of analyte; right: absence of analyte.

For small molecules (such as unbound fluorescein labelled low-molecular-weight antigen) with fast Brownian rotation in solution, FP is low, while for larger molecules (such as tracer bound to antibody) the values are higher. If the analyte is not present in the sample, the tracer will be bound to antibody and the FP will be high. If the analyte concentration in the sample is significantly higher than the concentration of the tracer, the antibody-binding site will be preferentially occupied by analyte and most of the tracer will be free in solution. The FP of the reaction mixture will then be lower [97]. The common FPIA procedure usually include addition of the sample to the diluted specific antibody solution, incubation and addition of the labelled analyte. For well-developed system, one-step single reagent assay is possible using a pre-equilibrated mixture of antibody and tracer [98,99].

As for other immunoreactions in solution, the equilibrium in the reaction mixture is reached in minutes or even seconds. As a consequence, the total time for detection of an analyte using FPIA may be only a few minutes. Because of its simplicity and speed, FPIA is readily automated and therefore suitable for high throughput screening in a variety of application areas.

FPIA have all advantages of homogeneous method such as simplicity, speed and high throughput. Specific FPIA advantage is tracer's stability on storage and benefit from the ability of fluorescein to fluoresce with high quantum yield. The small size of fluorescent label retains the immunoreactivity of the antigen. The method is less sensitive to the temperature changing than enzymatic reactions, but still sensitive enough because of temperature influence on the solution viscosity and so molecules and complexes rotation. FPIA outcomes sensitive to the presence of organic solvents (for example, extragents) because it also influenced solution viscosity and so rotation speed. The effective detection limit of FPIA typically does not compare with that of ELISA. FPIA is susceptible to interference from presence of endogenous fluorophores in samples, and from tracer binding to sample matrix components.

Application of FPIA is mostly limited to low-molecular-weight analytes. High-molecular-weight analytes are problematic for FPIA detection for two reasons. First, binding of a large antigen to an antibody, which may be of similar mass, produces a smaller relative change in the rotation rate and thus a considerably smaller change in polarization. Second, fluorescein molecules have excited state lifetime of 4.5 ns, which is too short to permit sensing of rotational moving of larger biomolecules.

Because of fluorescein dye's intrinsic spectral properties, light-scattering phenomena can interfere with these signals owing to the relatively small Stokes' shift (~30 nm). Alternative labels with improved characteristics such as long-wavelength organic dye, Nile blue [100], complexes of rhenium, osmium and ruthenium [101] or fluorescent QDs [102]. Introduction if new labels is partially restricted because of FPIA equipment for clinical laboratories is adjusted for fluorescein spectral properties.

Of course, it is impossible to evaluate FPIA outcomes by naked eye. Initially, the method was strictly limited because of the primitive state of commercial fluorimetres and the requirement for two separate measurements differing by a 90 degrees rotation of a polarizing lens. Fully automated polarization fluorimetre was developed and distribute by Abbott Laboratories (USA) in 1980, polarization units are available for several bench top and hand-handled fluorimetres. Application of FPIA extended from clinical laboratories (therapeutic drug monitoring and abused drug screening tests) to food control area (pesticides, mycotoxin detection) and environmental monitoring. FPIA application developed in the past few decades in many laboratories is reviewed in Refs [97,103–105]. Basic principles of fluorescence polarization/anisotropy phenomena and their application in diagnostics and imaging are detailed in Ref. [106].

3.3 Fluorescence Resonance Energy Transfer

Förster fluorescence resonance energy transfer (FRET) involves the nonradiative transfer of excitation energy from an excited donor fluorophore, D, (after absorption of a high energy photon) to a ground-state acceptor fluorophore, A, brought in close proximity, which can radiatively emit a low energy photon [107,108]. FRET processes are driven by dipole–dipole interactions and depend on the degree of spectral overlap between donor photoluminescence and acceptor absorption, and on the sixth power of the separation distance between the donor and acceptor pair (components of FRET nanoplatform). When both are fluorescent, energy transfer manifests as a decrease in donor fluorescence intensity and shortening of its exciton lifetime; this in turn coupled with enhancement of the acceptor longer wavelength fluorescence and increase of its exciton lifetime.

If only one component of the pair is luminescent, so only changing of one parameter could be detected. The rate of energy transfer is inversely dependent on the sixth power of the distance between the donor and acceptor and is directly related to the spectral overlap of the donor emission and acceptor absorption spectra. Distance at which FRET could be realized and posed distance is around 10 nm depending on the donor and acceptor nature. The power of this technique derives from its high intrinsic sensitivity to small changes (0.5–10 nm) in the separation distance and orientation between the donor and acceptor dipoles, which has led to the characterization of FRET in several reports as a 'spectroscopic ruler' [109]. The useful list of R_0 values (Förster transfer difference at which the energy transfer rate is equal to donor fluorescence decay rate) for over 70 donor–acceptor pairs for organic compounds are presented in Ref. [110]; these values are in the interval of 2–7 nm. This distance makes it possible to use FRET nanoplatform for biosensing not only in samples like serum, blood, urine, food extracts, but also inside living cells [111].

FRET immunoassay could be applicable for both low- and high-molecular-weight analytes in competitive and noncompetitive formats, respectively. Competitive FRET immunoassays (Fig. 18) constructed by conjugating a one component of donor–acceptor pair to the antigen and another to the antibody. If acceptor is a low-molecular-weight molecule, multiple acceptors are attached to the antibody to assure that there will be at least one acceptor at an efficient energy transfer distance within the immune complex. In competitive immunoassays, the presence of an antigen causes a reduction in the rate of decrease in donor fluorescence by competing for antibody-binding sites. In sandwich noncompetitive FRET assays (Fig. 19), donor and acceptor are

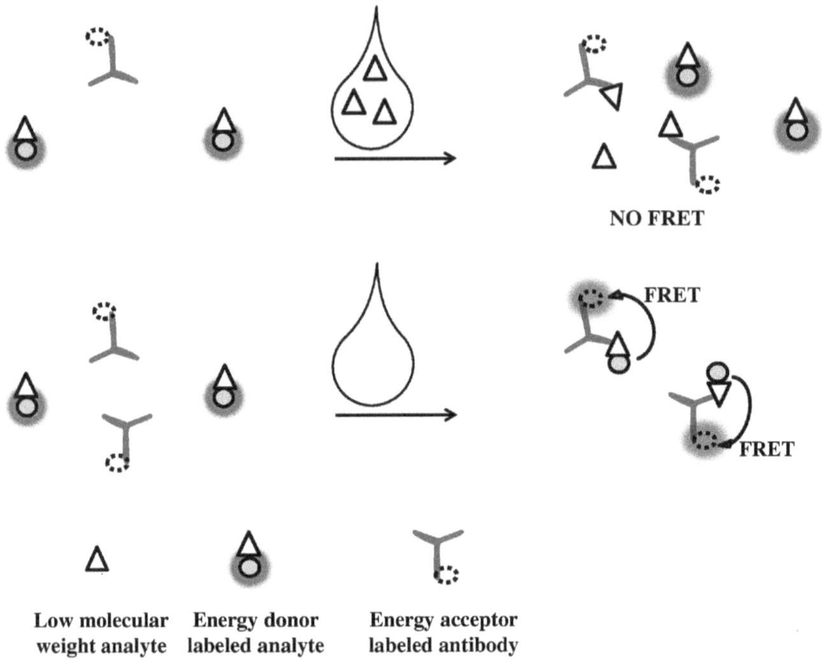

FIGURE 18 Competitive Förster fluorescence resonance energy transfer immunoassay.

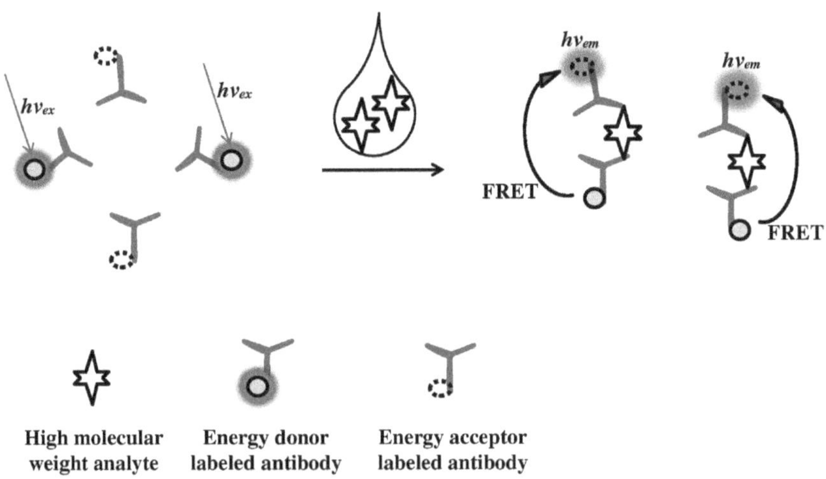

FIGURE 19 Noncompetitive sandwich Förster fluorescence resonance energy transfer immunoassay.

attached on each of the antibodies. If components of FRET pair are low-molecular-weight molecules, attachment of multiple molecules will increase FRET effectivity. Usually, the assays are carried out by following the rate of change in fluorescence intensity (or lifetime) during the initial phase of immunochemical binding. This approach reduces interference from the sample fluorescence, which normally does not change over the course of the assay. In sandwich immunoassays, increasing concentrations of antigen accelerate quenching of the donor emission due to FRET [89]. Originally FRET immunoassays were developed to avoid the need for measuring fluorescence polarization by the primitive methods available at that time [112,113]. Only a simple fluorimetre is needed, and the method is applicable to both small and large molecules [89,114].

The first commercial use of FRET immunoassays was in Syva's Advance immunochemistry system in the 1980s for serum cardiac glycoside digoxin assay. The new test, designed to run on Syva's Advance fluorometric instrument system, 'is the only digoxin assay now available that does not require time-consuming centrifugation, precipitation or separation by the lab technician', said Syva. With the Advance system, laboratories can decrease processing time for the test both on a volume basis, and by reducing the routine procedural steps required by a lab technician from seven (for an enzyme immunoassay), to four, according to the company. In this system, the reduction in fluorescence of the donor was monitored rather than the theoretically more sensitive appearance of fluorescence of the acceptor. This was necessary because of the difficulty in finding a fluorescent acceptor that is not directly excited by the light used to excite the donor. Nevertheless, by using phycoerythrin (a highly fluorescent red protein-pigment complex, isolated from red algae found in San Francisco and Monterey Bays, with an extinction coefficient near $2 \times 10^6\ M^{-1}cm^{-1}$) as the donor, sufficient sensitivity could be obtained for a homogeneous digoxin assay that could quantitate 500 pM of the drug in serum (25 pM in the assay medium) [89].

A crucial aspect of FRET development involves optimizing energy donor and acceptor to function in concert with desired recognition elements. Organic dyes are available to accomplish this, but have functional limitations such as pH dependence, susceptibility to photobleaching, and narrow excitation bands coupled with broad emission bands, all of which can compromise FRET performance [115]. To be perspective, components of FRET nanoplatform, both donor and acceptor should provide physical, chemical and photostability, maximal overlapping of donor emission spectrum and acceptor excitation spectrum, energy of the exited level of the donor should be higher or equal to the level of exited state of the acceptor. To make FRET most effective, ideal donor characteristics include high quantum yield, sufficient lifetime and broad excitation band coupled with narrow emission band. Ideal acceptor should have high absorption coefficient. High acceptor quantum yield allows to monitor FRET using an increase of acceptor photoluminescence. Presence of the two channels of signal monitoring (luminescence of donor and acceptor) makes it possible to use ratiometric approach and evaluate analytical signal as ratio of the two [111]. This approach allows to minimize influence of sample homogeneity and transmission fluctuation.

Development of nanoparticle-based structures has circumvented this problem through the development of specialized acceptors and donors. For more detailed information about properties and application of label, see in "Chapter 4, Labels for optical immunotests by Goryacheva" [130]. Here only FRET-related properties are briefly discussed.

Gold nanoparticles (AuNPs) have been found to be highly efficient FRET acceptors that permit direct monitoring of donor fluorescence without interfering acceptor auto-fluorescence [89,116–118]. Through FRET mechanism, AuNPs are effective quenchers for dye molecules [119,120], QDs [121], upconverting luminescent nanoparticles [122], graphene nanostructures [123]. Carbon-based structures could fulfil both donor and acceptor roles [123,124].

The use of QDs as an energy donor has recently received considerable attention. The properties of QDs can enhance energy transfer efficiencies, facilitate the design of donor–acceptor systems, simplify quantitative measurements, and enable multiplexing. There are three properties of QDs that are particularly advantageous from the standpoint of a donor for FRET:

1. they have similar but nonidentical broad absorption spectra and large molar absorption coefficients;
2. they have nontrivial and (bio)chemically accessible surface area [125].
3. they have narrow, size-tunable photoluminescence with good quantum yields. This advantage of QD donors is particularly useful in multiplexed FRET configurations – that is, the simultaneous interrogation of two or more donor–acceptor pairs in the same system. Thus, the use of dark quenchers as acceptors for QD donors can provide the opportunity for highly multiplexed FRET [126].

As acceptors, QDs are not so successful [127]. The broad absorption of QDs results in very efficient and unavoidable direct excitation – regardless of excitation wavelength. The use of QDs as acceptors has been demonstrated using other QDs [128] or luminescent lanthanide complexes as donors [129]. In these systems, the donor lifetime is comparable to, or longer than, the QD acceptor lifetime [125].

The use of rare earth chelates allows to develop of time-resolved FRET immunoassays, TR-FRET, provided a major improvement in fluorescence immunoassays. Rare earth chelates that serve as the donor are particularly useful for this purpose. These labels have long fluorescent lifetimes in the order of milliseconds whereas the fluorescent emission of most fluorescent substances decays in less than a microsecond [89].

ACKNOWLEDGEMENTS

The work was supported by Russian Science foundation (project 14-13-00229).

REFERENCES

[1] B.B. Dzantiev, N.A. Byzova, A.E. Urusov, A.V. Zherdev, Immunochromatographic methods in food analysis, Trends Anal. Chem. 55 (2014) 81–93.
[2] The Immunoassay Handbook, in: D. Wild (Ed.), Theory and Applications of Ligand Binding, ELISA Related Techniques, 2013, p. 1036.
[3] J.R. Crowther, The ELISA Guidebook, second ed., Humana Press, 2009, 566 pp.
[4] J.R. Crowther, ELISA: Theory and Practice, Springer Science & Business Media, 1995, 223 pp.
[5] R. van Dorp, P.G. Kok, E. Marani, M.E. Boon, L.P. Kok, ELISA incubation times can be reduced by 2.45-GHz microwaves, J. Clin. Lab Immunol. 34 (1991) 87–96.
[6] I.Y. Goryacheva, S. De Saeger, S.A. Eremin, C. Van Peteghem, Immunochemical methods for rapid mycotoxin detection: evolution from single to multiple analyte screening. A review, Food Addit. Contam. 24 (2007) 1169–1183.
[7] G.A. Posthuma-Trumpie, J. Korf, A. van Amerongen, Lateral flow (immuno)assay: its strengths, weaknesses, opportunities and threats. A literature survey, Anal. Bioanal. Chem. 393 (2009) 569–582.
[8] M.G. Weller, Immunochromatographic techniques – a critical review, Fresenius J. Anal.Chem. 366 (2000) 635–645.
[9] L.J. Janis, F.E. Regnier, Immunological-chromatographic analysis, J. Chromatogr. 444 (1988) 1–11.
[10] A.C. Moser, D.S. Hage, Immunoaffinity chromatography: an introduction to applications and recent developments, Bioanalysis 2 (2010) 769–790.
[11] R.C. Wong, H.Y. Tse, Lateral Flow Immunoassay, Humana Press, New York, 2009.
[12] B. O'Farrell, Lateral flow immunoassay systems: evolution from the current state of the art to the next generation of highly sensitive, quantitative rapid assays, in: D. Wild (Ed.), The Immunoassay Handbook Theory and Applications of Ligand Binding, ELISA and Related Techniques, fourth ed., Elsevier, 2013.
[13] J. Hu, S.Q. Wang, L. Wang, F. Li, B. Pingguan-Murphy, T. Jian Lu, F. Xu, Advances in paper-based point-of-care diagnostics, Biosens. Bioelectron. 54 (2014) 585–597.
[14] J.H. Cho, S.H. Paek, Semiquantitative, bar code version of immunochromatographic assay system for human serum albumin as model analyte, Biotechnol. Bioeng. 75 (2001) 725–732.
[15] D. Zhang, P. Li, Q. Zhang, R. Li, W. Zhang, X. Ding, C.M. Li, A naked-eye based strategy for semiquantitative immunochromatographic assay, Anal. Chim. Acta 740 (2012) 74–79.
[16] A.Y. Kolosova, L. Sibanda, F. Dumoulin, J. Lewis, E. Duveiller, C. Van Peteghem, S. De Saeger, Lateral-flow colloidal gold-based immunoassay for the rapid detection of deoxynivalenol with two indicator ranges, Anal. Chim. Acta 616 (2008) 235–244.
[17] C. Fang, Z. Chen, L. Li, J. Xia, Barcode lateral flow immunochromatographic strip for prostate acid phosphatase determination, J. Pharm. Biomed. Anal. 56 (2011) 1035–1040.
[18] A.Y. Kolosova, S. De Saeger, L. Sibanda, R. Verheijen, C. Van Peteghem, Development of a colloidal gold-based lateral-flow immunoassay for the rapid simultaneous detection of zearalenone and deoxynivalenol, Anal. Bioanal. Chem. 389 (2007) 2103–2107.

[19] W.B. Shim, B.B. Dzantiev, S.A. Eremin, D.H. Chung, One-step simultaneous immunochromatographic strip test for multianalysis of ochratoxin A and zearalenone, J. Microbiol. Biotechnol. 19 (2009) 83–92.
[20] C. Zhang, Y. Zhang, S. Wang, Development of multianalyte flow-through and lateral-flow assays using gold particles and HRP as tracers for the rapid determination of carbaryl and endosulfan in agricultural products, J. Agric. Food Chem. 54 (2006) 2502–2507.
[21] Y.R. Guo, S.Y. Liu, W.J. Gui, G.N. Zhu, Gold immunochromatographic assay for simultaneous detection of carbofuran and triazophos in water samples, Anal. Biochem. 389 (2009) 32–39.
[22] C. Fernandez-Sanchez, C.J. McNeil, K. Rawson, O. Nilsson, H.Y. Leung, V. Gnanapragasam, One-step immunostrip test for the simultaneous detection of free and total prostate specific antigen in serum, J. Immunol. Methods 307 (2005) 1–12.
[23] W. Leung, C.P.Y. Chan, M. Leung, K. Lehmann, I. Renneberg, M. Lehmann, A. Hempel, J.F.C. Glatz, R. Renneberg, Novel "digital-style" rapid test simultaneously detecting heart attack and predicting cardiovascular disease risk, Anal. Lett. 38 (2005) 423–439.
[24] Y. Oku, K. Kamiya, H. Kamiya, Y. Shibahara, T. Ii, Y. Uesaka, Development of oligonucleotide lateral-flow immunoassay for multi-parameter detection, J. Immunol. Methods 258 (2001) 73–84.
[25] J.V. Samsonova, V.A. Safronova, A.P. Osipov, Pretreatment-free lateral flow enzyme immunoassay for progesterone detection in whole cows' milk, Talanta 132 (2015) 685–689.
[26] R.F. Zuk, V.K. Ginsberg, T. Houts, J. Rabbie, H. Merrick, E.F. Ullman, M.M. Fischer, S.N. Sizto, D.J. Litman, Enzyme immunochromatography–a quantitative immunoassay requiring no instrumentation, Clin. Chem. 31 (1985) 1144–1150.
[27] H. Watanabe, A. Satake, M. Matsumoto, Y. Kido, A. Tsuji, K. Ito, M. Maeda, The home of premier fundamental discoveries, inventions and applications in the analytical and bioanalytical sciences, Analyst 123 (1998) 2573–2578.
[28] C. Parolo, A. de la Escosura-Muniz, A. Merkoci, Enhanced lateral flow immunoassay using gold nanoparticles loaded with enzymes, Biosens. Bioelectron. 40 (2013) 412–416.
[29] J.H. Cho, E.H. Paek, I.H. Cho, S.H. Paek, An enzyme immunoanalytical system based on sequential cross-flow chromatography, Anal. Chem. 77 (2005) 4091–4097.
[30] A.N. Kawde, X. Mao, H. Xu, Q. Zeng, Y. He, G. Liu, Moving enzyme-linked immunosorbent assay to the point-of-care dry reagent strip biosensors, Am. J. Biomed. Sci. 2 (2010) 23–32.
[31] T. Ono, M. Kawamura, S. Arao, H. Nariuchi, A highly sensitive quantitative immunochromatography assay for antigen-specific IgE, J. Immunol. Methods 272 (2003) 211–218.
[32] D. Du, J. Wang, L. Wang, D. Lu, Y. Lin, Integrated lateral flow test strip with electrochemical sensor for quantification of phosphorylated cholinesterase: biomarker of exposure to organophosphorus agents, Anal. Chem. 84 (2012) 1380–1385.
[33] I.H. Cho, E.H. Paek, Y.K. Kim, J.H. Kim, S.H. Paek, Chemiluminometric enzyme-linked immunosorbent assays (ELISA)-on-a-chip biosensor based on cross-flow chromatography, Anal. Chim. Acta 632 (2009) 247–255.
[34] H.A. Joung, Y.K. Oh, M.G. Kim, An automatic enzyme immunoassay based on a chemiluminescent lateral flow immunosensor, Biosens. Bioelectron. 53 (2014) 330–335.
[35] K. Inoue, P. Ferrante, Y. Hirano, T. Yasukawa, H. Shiku, T. Matsue, A competitive immunochromatographic assay for testosterone based on electrochemical detection, Talanta 73 (2007) 886–892.

[36] I.Y. Goryacheva, P. Lenain, S. De Saeger, Nanosized labels for rapid immunotests: new trends and developments, Trends Anal. Chem. 46 (2013) 30–46.

[37] J.H. Leuvering, P.J. Thal, M. van der Waart, A.H. Schuurs, Sol particle immunoassay (SPIA), J. Immunoassay Immunochem 1 (1980) 77–91.

[38] Y. Al-Yousif, J. Anderson, C. Chard-Bergstrom, S. Kapil, Development, evaluation, and application of lateral-flow immunoassay (immunochromatography) for detection of rotavirus in bovine fecal samples, Clin. Diagn. Lab. Immunol. 9 (2002) 723–724.

[39] N. Kusano, K. Hirashima, M. Kuwahara, K. Narahara, T. Imamura, T. Mimori, K. Nakahira, K. Torii, Immunochromatographic assay for simple and rapid detection of Satsuma dwarf virus and related viruses using monoclonal antibodies, J. Gen. Plant Pathol. 73 (2007) 66–71.

[40] R. Verheijen, P. Stouten, G. Cazemier, W. Haasnoot, Development of a one step striptest for the detective of sulfadomidine residues, Analyst 123 (1998) 2437–2441.

[41] G.C. Gussenhoven, M.A. van der Hoorn, M.G. Goris, W.J. Terpstra, R.A. Hartskeerl, B.W. Mol, C.W. van Ingen, H.L. Smits, LEPTO dipstick, a dipstick assay for detection of Leptospira-specific immunoglobulin M antibodies in human sera, J. Clin. Microbiol. 35 (1997) 92–97.

[42] S. Lou, C. Patel, S. Ching, J. Gordon, One-step competitive immunochromatographic assay for semiquantitative determination of lipoprotein(a) in plasma, Clin. Chem. 39 (1993) 619–624.

[43] M. Lonnberg, J. Carlsson, Quantitative detection in the attomole range for immunochromatographic tests by means of a flatbed scanner, Anal. Biochem. 293 (2001) 224–231.

[44] N.V. Zaytseva, R.A. Montagna, E.M. Lee, A.J. Baeumner, Multi-analyte single-membrane biosensor for the serotype-specific detection of Dengue virus, Anal. Bioanal. Chem. 380 (2004) 46–53.

[45] J.A. Ho, M.R. Huang, Application of a liposomal bioluminescent label in the development of a flow injection immunoanalytical system, Anal. Chem. 77 (2005) 3431–3436.

[46] J.A.A. Ho, R.D. Wauchope, A strip liposome immunoassay for aflatoxin B1, Anal. Chem. 74 (2002) 1493–1496.

[47] E.R. Goldman, A.R. Clapp, G.P. Anderson, H.T. Uyeda, J.M. Mauro, I.L. Medintz, H. Mattoussi, Multiplexed toxin analysis using four colors of quantum dot fluororeagents, Anal. Chem. 76 (2004) 684–688.

[48] M. Zuiderwijk, H.J. Tanke, R.S. Niedbala, P.L. Corstjens, An amplification-free hybridization-based DNA assay to detect *Streptococcus pneumoniae* utilizing the up-converting phosphor technology, Clin. Biochem. 36 (2003) 401–403.

[49] P. Corstjens, M. Zuiderwijk, A. Brink, S. Li, H. Feindt, R.S. Niedbala, H. Tanke, Use of up-converting phosphor reporters in lateral-flow assays to detect specific nucleic acid sequences: a rapid, sensitive DNA test to identify human papillomavirus type 16 infection, Clin. Chem. 47 (2001) 1885–1893.

[50] S.S. Deshpande, Enzyme Immunoassays: From Concept to Product Development, Chapman & Hall, New York, 1996.

[51] M.P. Laitinen, M. Vuento, Affinity immunosensor for milk progesterone: identification of critical parameters, Biosens. Bioelectron. 11 (1996) 1207–1214.

[52] M. O'Keeffe, P. Crabbe, M. Salden, J. Wichers, C. van Peteghem, F. Kohen, G. Pieraccini, G. Moneti, Preliminary evaluation of a lateral flow immunoassay device for screening urine samples for the presence of sulphamethazine, J. Immunol. Methods 278 (2003) 117–126.

[53] S.Z. Qian, H.H. Bau, Analysis of lateral flow biodetectors: competitive format, Anal. Biochem. 326 (2004) 211–224.
[54] J.Y. Lee, Y.A. Kim, M.Y. Kim, Y.T. Lee, B.D. Hammock, H.S. Lee, Importance of membrane selection in the development of immunochromatographic assays for low-molecular weight compounds, Anal. Chim. Acta 757 (2012) 69–74.
[55] S. Wang, Y. Quan, N. Lee, I.R. Kennedy, Rapid determination of fumonisin B1 in food samples by enzyme-linked immunosorbent assay and colloidal gold immunoassay, J. Agric. Food Chem. 54 (2006) 2491–2495.
[56] S. Wang, C. Zhang, J. Wang, Y. Zhang, Development of colloidal gold-based flow-through and lateral-flow immunoassays for the rapid detection of the insecticide carbaryl, Anal. Chim. Acta 546 (2005) 161–166.
[57] X. Xiang, W. Tianping, T. Zhigang, Development of a rapid, sensitive, dye immunoassay for schistosomiasis diagnosis: a colloidal dye immunofiltration assay, J. Immunol. Met. 280 (2003) 49–57.
[58] C. Paepens, S. De Saeger, L. Sibanda, I. Barna-Vetro, I. Leglise, F. Van Hove, C. Van Peteghem, A flow-through enzyme immunoassay for the screening of fumonisins in maize, Anal. Chim. Acta 523 (2004) 229–235.
[59] S. De Saeger, L. Sibanda, A. Desmet, C. Van Peteghem, A collaborative study to validate novel field immunoassay kits for rapid mycotoxin detection, Int. J. Food Microbiol. 75 (2002) 135–142.
[60] L. Sibanda, S. De Saeger, C. Van Peteghem, Development of a portable field immunoassay for the detection of aflatoxin M in milk, Int. J. Food Microbiol. 48 (1999) 203–209.
[61] P. Crabbe, C. Van Peteghem, Rapid and sensitive screening of sulfamethazine in porcine urine with an enzyme-linked immunosorbent assay and a field-portable immunofiltration assay, J. Food Prot. 65 (2002) 820–827.
[62] E. Schneider, E. Usleber, E. Maertlbauer, Rapid detection of Fumonisin B1 in corn-based food by competitive direct dipstick enzyme immunoassay/enzyme-linked immunofiltration assay with integrated negative control reaction, J. Agric. Food Chem. 43 (1995) 2548–2552.
[63] R. Collin, E. Schneider, L. Briggs, N. Towers, Development of immunodiagnostic field tests for the detection of the mycotoxin, sporidesmin A, Food Agric. Immunol. 10 (1998) 91–104.
[64] O. Buyuktanir, O. Genc, O. Celebi, N. Yurdusev, Rapid immunofiltration assay as a field diagnostic tool for ovine brucellosis, J. Immunoass. Immunochem. 33 (2012) 35–47.
[65] O. Genc, O. Buyuktanır, N. Yurdusev, Development of an individual rapid test based on enzymatic immunofiltration assay for detection of Anti-Brucella abortus antibody in bovine sera, J. Vet. Diagn. Invest. 23 (2011) 49–56.
[66] A. Pal, T.K. Dhar, An analytical device for on-site immunoassay. Demonstration of its applicability in semiquantitative detection of aflatoxin B1 in a batch of samples with ultrahigh sensitivity, Anal. Chem. 76 (2004) 98–104.
[67] A. Pal, D. Acharya, D. Saha, T.K. Dhar, Development of a membrane-based immunofiltration assay for the detection of T-2 toxin, Anal. Chem. 76 (2004) 4237–4240.
[68] A. Pal, D. Acharya, D. Saha, D. Roy, T.K. Dhar, In situ sample cleanup during immunoassay: a simple method for rapid detection of aflatoxin B1 in food samples, J. Food Prot. 68 (2005) 2169–2177.
[69] D. Saha, D. Acharya, T.K. Dhar, Method for homogeneous spotting of antibodies on membranes: application to the sensitive detection of ochratoxin A, Anal. Bioanal. Chem. 385 (2006) 847–854.

[70] D. Saha, D. Acharya, D. Roy, D. Shrestha, T.K. Dhar, Simultaneous enzyme immunoassay for the screening of aflatoxin B1 and ochratoxin A in chili samples, Anal. Chim. Acta 584 (2007) 343–349.

[71] E. Schneider, V. Curtui, C. Seidler, R. Dietrich, E. Usleber, E. Maertlbauer, Rapid methods for deoxynivalenol and other trichothecenes, Toxicol. Lett. 153 (2004) 113–121.

[72] L. Sibanda, S. De Saeger, I. Barna-Vetro, C. Van Peteghem, Development of a solid-phase cleanup and portable rapid flow-through enzyme immunoassay for the detection of ochratoxin A in roasted coffee, J. Agric. Food Chem. 50 (2002) 6964–6967.

[73] I.Y. Goryacheva, S. De Saeger, M. Lobeau, S.A. Eremin, I. Barna-Vetró, C. Van Peteghem, Approach for ochratoxin A fast screening in spices using clean-up tandem immunoassay columns with confirmation by high performance liquid chromatography–tandem mass spectrometry (HPLC–MS/MS), Anal. Chim. Acta 577 (2006) 38–45.

[74] I.Y. Goryacheva, N.V. Beloglazova, S.A. Eremin, D.A. Mikhirev, R. Niessner, D. Knopp, Gel-based immunoassay for non-instrumental detection of pyrene in water samples, Talanta 75 (2008) 517–522.

[75] A. Lucht, P. Formenty, H. Feldmann, M. Götz, E. Leroy, P. Bataboukila, A. Grolla, F. Feldmann, T. Wittmann, P. Campbell, C. Atsangandoko, P. Boumandoki, E.J. Finke, P. Miethe, S. Becker, R. Grunow, Development of an immunofiltration-based antigen-detection assay for rapid diagnosis of Ebola virus infection, J. Infect. Dis. 196 (Suppl. 2) (2007) S184–S192. http://jid.oxfordjournals.org/content/196/Supplement_2/S184.short-aff-4.

[76] V.V. Goftman, N.V. Beloglazova, E. Njumbe Ediage, S. De Saeger, R. Dietrich, E. Märtlbauer, I.Y. Goryacheva, Rapid immunochemical tests for qualitative and quantitative determination of T-2 and HT-2 toxins, Anal. Methods 4 (2012) 4244–4249.

[77] M. Lobeau, S. De Saeger, L. Sibanda, I. Barna-Vetró, C. Van Peteghem, Development of a new clean-up tandem assay column for the detection of ochratoxin A in roasted coffee, Anal. Chim. Acta 538 (2005) 57–61.

[78] M. Lobeau, S. De Saeger, L. Sibanda, I. Barna-Vetró, C. Van Peteghem, Application and validation of a clean-up tandem assay column for screening ochratoxin A in cocoa powder, Food Addit. Contam. 24 (2007) 398–405.

[79] I.Y. Goryacheva, S. De Saeger, I.S. Nesterenko, S.A. Eremin, C. Van Peteghem, Rapid all-in-one three-step immunoassay for non-instrumental detection of ochratoxin A in high coloured herbs and spices, Talanta 72 (2007) 1230–1234.

[80] I.Y. Goryacheva, S. De Saeger, B. Delmulle, M. Lobeau, S.A. Eremin, I. Barna-Vetro, C. Van Peteghem, Simultaneous noninstrumental detection of aflatoxin B1 and ochratoxin A using a clean-up tandem immunoassay column, Anal. Chim. Acta 590 (2007) 118–124.

[81] I.Y. Goryacheva, T.Y. Rusanova, N.A. Burmistrova, S. De Saeger, Immunochemical methods for the determination of mycotoxins, J. Anal. Chem. 64 (2010) 768–785.

[82] N.V. Beloglazova, I.Y. Goryacheva, R. Niessner, D. Knopp, A comparison of horseradish peroxidase, gold nanoparticles and qantum dots as labels in non-instrumental gel-based immunoassay, Microchim. Acta 175 (2011) 361–367.

[83] N.V. Beloglazova, I.Y. Goryacheva, D.A. Mikhirev, S. De Saeger, R. Niessner, D. Knopp, New immunochemically-based field test for monitoring benzo[a]pyrene in aqueous samples, Anal. Sci. 24 (2008) 1613–1617.

[84] T.Y. Rusanova, N.V. Beloglazova, I.Y. Goryacheva, M. Lobeau, C. Van Peteghem, S. De Saeger, Non-instrumental immunochemical tests for rapid ochratoxin A detection in red wine, Anal. Chim. Acta 653 (2009) 97–102.

[85] I.Y. Goryacheva, M.A. Karagusheva, C. Van Peteghem, L. Sibanda, S. De Saeger, Gel-based immunoassay for non-instrumental screening of aflatoxin M1 in milk, Food Control 20 (2009) 802–806.
[86] E.Y. Basova, I.Y. Goryacheva, T.Y. Rusanova, N.A. Burmistrova, R. Dietrich, E. Märtlbauer, C. Detavernier, C. Van Peteghem, S. De Saeger, An immunochemical test for rapid screening of zearalenone and T-2 toxin, Anal. Bioanal. Chem. 397 (2010) 55–62.
[87] R. Kraus, Wien. Klin. Wochenschr 10 (1897) 736. As quoted by Kabat, E.A. & Mayer, M. Experimental Immunochemistry. (Charles C. Thomas, Springfield, IL, 1961).
[88] K. Meyer, Über hämagglutininvermehrung und hämagglutinationfordernde wirkung bei menschlichen seren, Z. Immun. exp. ther. 34 (1922) 229.
[89] E.F. Ullman, Homogeneous immunoassays, in: D. Wild (Ed.), The Immunoassay Handbook: Theory and Applications of Ligand Binding, ELISA Related Techniques, 2013, p. 1036.
[90] J. Stanley, Essentials of immunology and serology, Delmar Cengage Learn. (2002) 448.
[91] A.A. Panackal, J.P. Dekker, M. Proschan, A. Beri, P.R. Williamson, Enzyme immunoassay versus latex agglutination cryptococcal antigen assays in adults with non-HIV-related cryptococcosis, J. Clin. Microbiol. 52 (2014) 4356–4358.
[92] N.E. Babady, J.E. Bestrom, D.J. Jespersen, M.F. Jones, E.M. Beito, M.J. Binnicker, N.L. Wengenack, Evaluation of three commercial latex agglutination kits and a commercial enzyme immunoassay for the detection of cryptococcal antigen, Med. Mycol. 47 (2009) 336–338.
[93] G.K. Von Schulthess, R.J. Cohen, G.B. Benedek, Laser light scattering spectroscopic immunoassay in the agglutination-inhibition mode for human chorionic gonadotropin (hCG) and human luteinizing hormone, Immunochemistry 13 (1976) 963–966.
[94] S. Gupta, P.K. Kilpatrick, E. Melvin, O.D. Velev, On-chip latex agglutination immunoassay readout by electrochemical impedance spectroscopy, Lab Chip 12 (2012) 4279–4286.
[95] W.B. Dandliker, G. Feigen, Quantification of the antigen-antibody reaction by the polarization of fluorescence, Biochem. Biophys. Res. Commun. 5 (1961) 299–304.
[96] W.B. Dandliker, R.J. Kelly, J. Dandliker, J. Farquhar, J. Levin, Fluorescence polarization immunoassay. Theory and experimental method, Immunochemistry 10 (1973) 219–227.
[97] D.S. Smith, S.A. Eremin, Fluorescence polarization immunoassays and related methods for simple, high-throughput screening of small molecules, Anal. Bioanal. Chem. 391 (2008) 1499–1507.
[98] M.J. Choi, J.R. Lee, S.A. Eremin, Development of single reagent for fluorescence polarization immunoassay of atrazine, Food Agric. Immunol. 14 (2002) 107–120.
[99] J.Y. Hong, M.J. Choi, Development of one-step fluorescence polarization immunoassay for progesterone, Biol. Pharm. Bull. 25 (2002) 1258–1262.
[100] M.L. Sanchez Martinez, M.P. Aguilar Caballos, A. Gomez Hens, Long-wavelength fluorescence polarization immunoassay: determination of amikacin on solid surface and gliadins in solution, Anal. Chem. 79 (2007) 7424–7430.
[101] X.Q. Guo, F.N. Castellano, L. Li, J.R. Lakowicz, Use of a long-lifetime Re(I) complex in fluorescence polarization immunoassays of high-molecular-weight analytes, Anal. Chem. 70 (1998) 632–637.
[102] J. Tian, L. Zhou, Y. Zhao, Y. Wang, Y. Peng, S. Zhao, Multiplexed detection of tumor markers with multicolor quantum dots based on fluorescence polarization immunoassay, Talanta 92 (2012) 72–77.
[103] J.C. Owicki, Fluorescence polarization and anisotropy in high throughput screening: perspectives and primer, J. Biomol. Screen 5 (2000) 297–306.

[104] M.E. Jolley, M.S. Nasir, The use of fluorescence polarization assays for the detection of infectious diseases, Comb. Chem. High. Throughput Screen. 6 (2003) 235–244.

[105] J.R. Sportsman, Fluorescence anisotropy in pharmacologic screening, Methods Enzymol. 361 (2003) 505–529.

[106] D.M. Jameson, J.A. Ross, Fluorescence polarization/anisotropy in diagnostics and imaging, Chem. Rev. 110 (2010) 2685–2708.

[107] T. Förster, Discuss. Faraday Soc. 27 (1959) 7–17.

[108] J.R. Lakowicz, Principles of Fluorescence Spectroscopy, second ed., Kluwer Academic, New York, 1999.

[109] A.R. Clapp, I.L. Medintz, H. Mattoussi, Förster resonance energy transfer investigations using quantum-dot fluorophores, Chem. Phys. Chem. 7 (2006) 47–57.

[110] P. Wu, L. Brand, Review: resonance energy transfer-methods and applications, Anal. Biochem. 21 (1994) 1–13.

[111] C. Yu, X. Li, F. Zeng, F. Zheng, S. Wu, Carbon-dot-based ratiometric fluorescent sensor for detecting hydrogen sulfide in aqueous media and inside live cells, Chem. Commun. 49 (2013) 403–405.

[112] E.F. Ullman, P.L. Khanna, Fluorescence excitation transfer immunoassay, Meth. Enzymol. 74 (1981) 28–60.

[113] E.F. Ullman, M. Schwarzberg, K. Rubenstein, Fluorescent excitation transfer immunoassay, J. Biol. Chem. 251 (1976) 4172–4178.

[114] R.M. Clegg, Fluorescence resonance energy transfer, Curr. Opin. Biotechnol. 6 (1995) 103–110.

[115] I.L. Medintz, A.R. Clapp, H. Mattoussi, E.R. Goldman, B. Fisher, J.M. Mauro, Self-assembled nanoscale biosensors based on quantum dot FRET donors, Nat. Mater. 2 (2003) 630–638.

[116] K. Saha, S.S. Agasti, C. Kim, X. Li, V.M. Rotello, Gold nanoparticles in chemical and biological sensing, Chem. Rev. 112 (2012) 2739–2779.

[117] P.C. Ray, G.K. Darbha, A. Ray, J. Walker, W. Hardy, Gold nanoparticle based FRET for DNA detection, Plasmonics 2 (2007) 173–183.

[118] D.A. Giljohann, D.S. Seferos, W.L. Daniel, M.D. Massich, P.C. Patel, C.A. Mirkin, Gold nanoparticles for biology and medicine, Angew. Chem. Int. Ed. 49 (2010) 3280–3294.

[119] E. Dulkeith, A.C. Morteani, T. Niedereichholz, T.A. Klar, J. Feldmann, S.A. Levi, F.C.J.M. van Veggel, D.N. Reinhoudt, M. Möller, D.I. Gittins, Fluorescence quenching of dye molecules near gold nanoparticles: radiative and nonradiative effects, Phys. Rev. Lett. 89 (2002) 203002.

[120] S. Mayilo, M.A. Kloster, M. Wunderlich, A. Lutich, T.A. Klar, A. Nichtl, K. Kurzinger, F.D. Stefani, Feldmann, Long-range fluorescence quenching by gold nanoparticles in a sandwich immunoassay for cardiac troponin T, Nano Lett. 9 (2009) 4558–4563.

[121] E. Oh, M.Y. Hong, D. Lee, S.H. Nam, H.C. Yoon, H.S. Kim, Inhibition assay of biomolecules based on fluorescence resonance energy transfer (FRET) between quantum dots and gold nanoparticles, J. Am. Chem. Soc. 127 (2005) 3270–3271.

[122] L. Wang, R. Yan, Z. Huo, L. Wang, J. Zeng, J. Bao, X. Wang, Q. Peng, Y. Li, Fluorescence resonant energy transfer biosensor based on upconversion-luminescent nanoparticles, Angew. Chem. Int. Ed. 44 (2005) 6054–6057.

[123] J. Shi, C. Chan, Y. Pang, W. Ye, F. Tian, J. Lyu, Y. Zhang, M. Yang, A fluorescence resonance energy transfer (FRET) biosensor based on graphene quantum dots (GQDs) and gold nanoparticles (AuNPs) for the detection of mecA gene sequence of *Staphylococcus aureus*, Biosens. Bioelectron. 67 (2015) 595–600.

[124] M. Liu, H. Zhao, X. Quan, S. Chen, X. Fan, Distance-independent quenching of quantum dots by nanoscale-graphene in self-assembled sandwich immunoassay, Chem. Commun. 46 (2010) 7909–7911.

[125] W.R. Algar, M. Massey, U.J. Krull, Semiconductor quantum dots and FRET, in: I. Medintz, N. Hildebrandt (Eds.), FRET – Förster Resonance Energy Transfer: From Theory to Applications, Wiley-VCH Verlag GmbH & Co, KGaA, 2014, 791 pp.

[126] A.R. Clapp, I.L. Medintz, H.T. Uyeda, B.R. Fisher, E.R. Goldman, M.G. Bawendi, H. Mattoussi, Quantum dot-based multiplexed fluorescence resonance energy transfer, J. Amer. Chem. Soc. 127 (2005) 18212–18221.

[127] A.R. Clapp, I.L. Medintz, B.R. Fisher, G.P. Anderson, H. Mattoussi, Can luminescent quantum dots be efficient energy acceptors with organic dye donors? J. Amer. Chem. Soc. 127 (2005) 1242–1250.

[128] A.L. Rogach, T.A. Klar, J.M. Lupton, A. Meijerink, J. Feldmann, Energy transfer with semiconductor nanocrystals, J. Mater. Chem. 19 (2009) 1208–1221.

[129] L.J. Charbonniere, N. Hildebrandt, R.F. Ziessel, H.G. Löhmannsröben, Lanthanides to quantum dots resonance energy transfer in timeresolved fluoro-immunoassays and luminescence microscopy, J. Amer. Chem. Soc. 128 (2006) 12800–12809.

[130] I.Y. Goryacheva, Labels for optical immunotests, in: I.Y. Goryacheva (Ed.), Rapid Immunotests for Clinical, Food and Environmental Applications vol. 72, 2016, pp. 79–132.

Chapter 4

Labels for Optical Immunotests

Chapter Outline

1. INTRODUCTION

In order to improve the performance of rapid tests, innovations have been required in each of the key components of the test construction and technologies. One of the main area bringing possibilities is improvement assay sensitivity and simplicity in the development of signal generation technologies. The signal generation elements involve both the label and the reader or interpretation method used. The majority of detection is still done visually either by eye or using optical readers, however, there is growing use of fluorescent, chemiluminescent, and magnetic field measurement systems [1].

The main tendency in the development of new labels for rapid methods could be outlined as follows: initially as biolabels in biochemical research, then as application in traditional immunoassay formats ('new-label-linked

Comprehensive Analytical Chemistry, Vol. 72. http://dx.doi.org/10.1016/bs.coac.2016.03.008

immunosorbent assay') and finally in the development of sensors and rapid tests. Another common tendency is a shift from the clinical area (point-of-care testing) towards the food and feed control area and finally to the field of environmental control. The immunoassay format changes in parallel: from a sandwich format for high-weight analytes of clinical interest to a competitive format for mostly low-weight analytes in ecological monitoring. In spite of substantial interest in in-field assay, the main problem restricting rapid immunotest application is its lack of sensitivity. To solve this problem two main approaches are used: the first is an improvement of the recognition system and assay optimization whereas the second is the development of new labels or label systems thereby improving analytical signal generation through higher brightness and/or reader optimization.

Several reviews were published in nearby areas. Data about lateral-flow immunoassay (LFIA) development were summarized in reviews of Posthuma-Trumpie et al. [2] and Dzantiev et al. [3], protein rapid tests in review of Axel Warsinke [4]. Application of nanoparticle labels in immunosensing with optical detection was reviewed by Matthias Seydack [5]. Progress in the development of different types of inorganic nanoparticles used as labels in immunoassay was reviewed by Cháfer-Pericas et al. in 2006 [6] and the nanoparticles' preparation methods and surface-functionalization procedures were focussed on to improve the sensitivity of the involved immunoassays. Application of nanoparticles [7] and quantum dot—based labels [8] for rapid tests was reviewed by Goryacheva et al. A review concerning the trends in fluorescence immunochromatography was published by Pyo and Yoo [9].

This chapter presents an overview of the labels for rapid immunochemical tests, the principles of signal generation and the features of applications. Both traditional (enzymes, gold nanoparticles) and relatively new labels (up-converting phosphors (UCP), quantum dots (QD)) are presented. Comparison of sensitivity provided by tests with different labels is presented, where possible.

Optimal labels should be stable (colloidal) in water, have uniform size, charge and shape, be easily conjugatable to biomolecules, be able to produce an intense analytical signal and show stability against aggregation/oxidation/precipitation/photobleaching and so on. during the test preparation and assay procedure. Most of these characteristics will be affected by the analytical system, the tested samples and other factors. Labels can therefore be comparatively evaluated only by experimental testing.

2. ENZYMES

Enzymes are very popular label both for new formats development and for practical applications. Enzymes are catalysts, which bring about the change in the substrate without itself changing chemically and can therefore keep on converting the substrate. The main advantages of enzymes are:

- sensitivity;
- biocompatibility;
- availability of techniques, protocols, practices and readers for their application.

Undoubtedly, enzyme labels are used more widely than any other type of labels, because a single molecule of enzyme, for example, horseradish peroxidase, may cause the conversion of 10^7 molecules of substrate per minute [10]. The two enzymes most commonly used in immunoassays are alkaline phosphatase and horseradish peroxidase. Both are quite stable when handled and stored properly, and both can be stored at 4°C for more than 6 months. They are commercially available as free enzymes and as various enzyme conjugates (enzyme-labelled antibodies, antigens, DNA, etc.) and are relatively inexpensive; variety of conjugation techniques are well established [11]. Besides, if it is not convenient to enzyme conjugate the desired detection reagent, a secondary detection reagent can be used. For example, if a custom-generated mouse monoclonal antibody is used as the primary detection reagent, a commercially available horseradish peroxidase-conjugated goat antimouse antibody can be used as a secondary detection reagent to generate analytical signal. The incubation temperature, buffer composition and pH, enzyme stability and requirement for cofactors should all be considered when selecting an enzyme. There are some differences between these two enzymes that should be considered when choosing one for an assay.

Horseradish peroxidase is a 44-kDa glycosylated hemoprotein. It consists of a protoporphyrin IX hemin prosthetic group and 308 amino acids, including four disulphide bridges and two calcium ions. It contains about 20% carbohydrate by weight. It has six lysine residues available for conjugation without any loss of enzyme activity. Horseradish peroxidase is an oxidoreductase that can be used with a wide variety of hydrogen donors to reduce hydrogen peroxide. This property has been utilized to generate coloured, fluorescent or luminescent derivatives, depending on the substrate used. It has a high catalytic rate, resulting in amplification of the signal. Due to its small size, it rarely causes steric hindrance problems with antibody/antigen complexes bound on a surface. Peroxidase is very inexpensive compared to alkaline phosphatase. The major disadvantage associated with peroxidase is that it is incompatible with many preservatives, such as sodium azide, that are used to reduce microbial contamination in many biological buffer solutions. Sodium azide, even in low concentrations, inactivates peroxidase activity. The pH range of HRP activity is 4.0–8.0 [10]. Several substrates, yielding either soluble or insoluble reaction products, are commercially available for peroxidase.

Alkaline phosphatase is a 140-kDa dimeric glycoprotein containing many free amino groups that can be used for conjugation. This enzyme catalyzes the hydrolysis of phosphate esters of primary alcohols, phenols and amines. The form used in immunoassays is purified from calf intestine and contains zinc. It is inhibited by orthophosphate, zinc chelators, borate, carbonate and urea. Activity is optimal at pH 9.5−10.5 [10]. The larger molecular size of alkaline phosphatase can cause steric hindrance issues due to closely packed antigen−antibody complexes. This can result in lower activity than expected for the estimated number of bound enzyme molecules.

Many other enzymes have been tried over the years in immunoassays (eg, acetate kinase, firefly luciferase, xanthine oxidase), with β-D-galactosidase, glucose oxidase and glucose 6-phosphate dehydrogenase being the only ones to find application in commercial products [10].

The main disadvantage of enzymes as labels is not providing possibility for direct detection, which means necessity for additional steps for signal development. While most test systems are based on rapid analyses with the minimum number of steps, enzymes need prolonged incubation with an appropriate substrate. This way, compared with labels that provide direct detection, enzyme-based immunoassays usually have some disadvantages [10,12]:

1. Enzyme-based immunoassays are susceptible to interference and changes in assay conditions during the signal generation stage.
2. Signal generation must be controlled, optimized and kept free from interferences, in much the same way as antibody−antigen incubation.
3. Enzyme−substrate incubation is sensitive to time, temperature and pH, and inhibitory substances should be absent.
4. Substrate is normally present in excess, except where its high concentration inhibits enzyme activity, as is the case with horseradish peroxidase and hydrogen peroxide.

Enzymes can be covalently linked to the antigen or antibody (depending on the format of the assay) to form a conjugate using a wide range of chemical techniques [10]. Enzyme labels may give rise to several different signal sources, depending on the substrate; but non-depending on the detection principle, enzyme application needs substrate to develop analytical signal. This means rapid enzyme−based tests could commonly not work in dry-chemistry format as, for example, lateral-flow immunostrips with nanoparticles as labels. At the same time comparative studies of enzyme and nanoparticle label showed that complication of the procedure gives gain in better sensitivity [13,14].

To additionally improve assay sensitivity, the common approach as to enhance amount of labels bound to each immunoreagent unit (antigen or antibody) which allows to enhance amount of enzymes per each analyte is used. For some enhanced labels, additional steps are prescribed to release enzymes from label internal volume before substrate application, such as liposome lysis, for example. For more details about liposomes (see Section 4.4). Other enhanced labels do not need any additional steps, such as commercial polyhorseradish peroxidase products (Thermo Fisher Scientific Inc. (USA), Senova GmbH (Germany)). Commercially available from Senova polyHRP products are supermolecular enzyme complexes containing up to 800 molecules of the enzyme. Enzyme molecules are cross-linked in the manufacture of PolyHRP by means of bifunctional linkers to form oligomers consisting of 20, 40 or 80 molecules. Preconcentration of enzyme immunocomplexes on the surface of nanoparticles, for example, magnetic, also allow to combine multiple enzymes in one label [15].

Parolo et al. [16] studied enhanced labels consisted of gold nanoparticles modified with horseradish peroxidase. It was shown that all three tested substrates (3,3′,5,5′-tetramethylbenzidine, 2,2′-azinobis[ethylbenzothiazoline-6-sulphonate] and 3,3′-diaminobenzidine with metal enhancer) produced a darker colour that enhanced the intensity of the previous red colour of the unmodified gold nanoparticles. In such very simple way it is gaining sensitivity (up to one order of magnitude). Presence of gold nanoparticles offers a dramatic visual enhancement of analytical signal of insoluble enzymatic catalytic product (red-coloured chromogen). Combining enzyme catalytic amplification (with 3-amino-9-ethylcarbazole (AEC)/H_2O_2 substrate solution) with unique optical properties of gold nanoparticles was capable of detecting of 0.01-pM target DNA without instrumentation [17].

Among rapid tests, rapid ELISA and flow-through immunoassay use enzymes as the main kind of labels; different variants of lateral-flow immunoassay also use enzyme or enzyme-containing labels. Variety of enzymes and related substrates provide different kinds of analytical signal evaluation: intensity of colour, fluorescence, chemiluminescence and electrochemiluminescence detection.

2.1 Colorimetric Assays

The simplest and most popular signal to measure is colour. Results could be estimated visually by naked eyes or quantified by different readers. The readers are kind of standard equipment to immunoassay-related methods. This method works well with horseradish peroxidase and 3,3′,5,5′-tetramethylbenzidine, 2,2′-azinobis(ethylbenzothiazoline-6-sulphonate) or *o*-phenylenediamine as substrates. Each of these substrates also requires the presence of hydrogen peroxide. Another popular enzyme is alkaline phosphatase in combination with *p*-nitrophenyl phosphate or indoxyl phosphate substrates. Horseradish peroxidase–based assays are capable of greater sensitivity than alkaline phosphatase with colorimetric substrates, as the signal intensity (under identical conditions) is an order of magnitude greater [10].

Colorimetric (or chromogenic) kind of detection supersedes radioisotopic labels. In theory, colorimetric assays are capable of sensitivity that is as good as that for radioimmunoassay, but this is seldom achieved in practice. Colorimetric assays are primarily limited by the lower and upper limits of spectrophotometric measurements; colorimetry sensitivity limited sensitivity of measuring the difference between two large light signals: incident and transmitted light [18]. Enzyme-based colorimetric immunoassay widely used for rapid ELISAs (Section 2.2.1), flow-through tests (Section 2.2.2) and different variations of dipsticks and ELISA-on-a-chip formats.

The signal generation stage involves the addition of substrate in buffer; during following incubation, the enzyme gradually converts a proportion of the substrate to its coloured endproduct. For convenient and rapid ELISAs at the end of the incubation, a stopping reagent is added that stops enzyme activity, for example by changing the pH [10]. For most of the nonmicrotitre plate—based methods, it is necessary to control incubation time very carefully, because no stopping reagent is usually used. The strength of the colour depends on the amount of enzyme present, which in a competitive assay is inversely proportional to the concentration of analyte in the original sample and directly proportional in a sandwich assay. To have an endpoint assay that provides reliable and consistent results, it is important that the timing of the reaction in each and every test be controlled as precisely as possible. Since enzyme—substrate reactions are kinetic, timing from the start to the end of the reaction strongly affects the final concentration of product developed.

Sensitivity of detection was compared for membrane-based competitive immunoassays using gold particles and horseradish peroxidase as tracers in flow-through and lateral-flow formats for multianalysis of carbaryl and endosulphan: 10 times enhancement in the visual detection limit using horseradish peroxidase label was obtained in the study [19].

For the analyte concentration semiquantification, Fung et al. [20]. used barcode principle with increasing number of coloured lines with increasing analyte concentrations. Analyte concentration can be evaluated by simply counting the number of lines. The dynamic range of the assay was 0—400 μM for creatinine [20] and 5—100 μM for glucose [21].

As enzyme application anyway required additional steps (comparing with normal nanoparticle—based lateral-flow strips), it is worth the effort to resort to various tricks for further sensitivity enhancement. A plastic chip that can perform immunoassays using an enzyme as signal generator was developed by incorporating an immunostrip into channels etched on the surfaces of the chip. To utilize an analytical concept of cross-flow chromatography, the chip consisted of two cross-flow channels in the horizontal and vertical directions. In the vertical channel, 2-mm wide immunostrip for cardiac troponin I was placed, which was identical to a conventional rapid test kit except for the utilization of an enzyme, horseradish peroxidase, as tracer [22].

To enhance sensitivity of strips, it was combined with an immunomagnetic separation and preconcentration using magnetic particles, coupled with specific antibody and enzyme. For detection of *Listeria monocytogenes* in food matrices, the immunomagnetic complex was added into sample for separation and preconcentration by magnetic nanoparticles (MNPs) containing the enzyme and directly applied to the assay system to induce an antigen–antibody reaction without any additional step [23]. The detection concept was demonstrated by performing an enzyme immunoassay on a microtitre well. Results show that the analytical performance of magnetic particles, coupled with specific antibody and enzyme is higher than that of the enzyme-labelled antibody, because higher amount of enzyme was bound to MNPs with large surface area. Results of the chromatographic analysis yield a limit of detection of 97 ± 19.5 CFU/mL in 2% milk sample [22].

2.2 Fluorescence Assays

Fluorescence (photoluminescence) signal measurement allows to reach sensitivity several orders of magnitude better than colorimetric assays using the same enzymes in combination with special substrates. There is a greater relative improvement in sensitivity with alkaline phosphatase than with horseradish peroxidase, and fluorescence immunoassay that utilizes this enzyme has a level of sensitivity approaching those based on horseradish peroxidase. β-galactosidase is also used frequently due to its greater sensitivity when used with a fluorogenic substrate. Additionally an application of fluorescence as signal registration channel widen the dynamic range of the assay by allowing very high readings to be accurately measured. Fluorescence is directly affected by temperature, polarity, pH and the dissolved oxygen content.

Substrates for fluorescence signal measurements should be converted in the presence of the enzyme to photoluminescent end products [10]. Stabilized substrates for all three of the top enzymes are commercially available. The resulting enzyme–substrate reaction product should have distinctly separate excitation and emission wavelengths (high Stokes shift). If the Stokes' shift is large, it is easier to measure the emitted light without interference from the incident light. The substrate itself should be nonfluorescent.

The most common substrate used in fluorescence immunoassay is 4-methylumbelliferyl phosphate with alkaline phosphatase as the enzyme label. 4-methylumbelliferyl phosphate is converted to 4-methylumbelliferone with an excitation wavelength of 360 nm and an emission wavelength of 440 nm. A suitable fluorogenic substrate for β-galactosidase is (4-methylumbelliferyl galactoside), which is converted to 4-methylumbelliferone (this is the same product that results from the conversion of 4-methylumbelliferyl phosphate, the substrate used with alkaline phosphatase). For horseradish peroxidase, two fluorogenic substrates are currently available: 3-(4-hydroxyphenyl) propionic acid and hydroxyphenylacetic acid. Both substrates require the addition of hydrogen peroxide in order to produce a fluorescent product. Currently, highly sensitive fluorescent methods using 3-(4-hydroxyphenyl) propionic acid is developed using an excitation wavelength of 320 nm and an emission wavelength of 404 nm.

Fluorescence is highly affected by temperature, polarity and pH; and for fluorophores with high lifetime, dissolved oxygen could decrease emission intensity through quenching of exciting states. Background fluorescence is naturally present in serum and plasma samples, caused by proteins at shorter wavelengths, and nicotinamide adenine dinucleotide and bilirubin at longer wavelengths. The peak fluorescence emission of plasma is around 350 nm (excitation at around 280 nm), with the weaker (but sometimes more critical) bilirubin peak at 520 nm. Quenching can occur when molecules in the sample absorb excitation light from the fluorometer or emitted light from the fluorophore [10].

2.3 Chemiluminescence Assays

Chemiluminescence is the phenomenon of light emission as a result of a chemical reaction. This means no excitation light (and so no related fluctuations) is involved. Unlike fluorescence measurement, there is no incident light, and the only signal emanates from the chemiluminescent molecules, which means the minimum background. Unlike radioactivity measurements, all the photons from the chemiluminescent compounds can be triggered in a very short time, much increasing the potential analytical signal. For example, the final data was acquired within 30 s after the addition of the enzyme substrate, which was faster than the detection time required when using a colorimetric substrate with the same tracer enzyme [15]. This explains the possibility to achieve levels of sensitivity several orders of magnitude better than radioisotopic and fluorometric immunoassays.

There are substrates that give rise to chemiluminescent end points for all of the commonly used enzyme labels, most notably horseradish peroxidase and alkaline phosphatase. The most usable chemiluminescent substrate for horseradish peroxidase is luminol, which is commercially available in a stabilized form. It is the most suited for clinical rapid tests. Commercially available luminol is provided with an enhancer (phenols, naphthols, aromatic amines or benzothiazoles) that acts as an enzyme protector and allows the reaction to proceed for several minutes without substantial loss in light output. pH must be stabilized at about 8.5 to allow both peroxidase activity (optimal at pH 5.5) and light emission (optimal at pH 12.0) to occur. Polyphenols are a class of horseradish peroxidase substrates that include pyrogallol, purpurogallin, gallic acid and umbelliferone. All polyphenols are known for their excellent signal to noise ratio and extremely rapid light decay. Polyphenol and acridine ester substrates can only be used in conjunction with luminescent detectors equipped to handle 'flash' reactions [10]. Commercially available substrate Lumigen TMA-6 is optimized for membrane-based assays. 3-(2′-spiroadamantane)-4-methyl-4-(3′-phosphoryloxyphenyl-1, 2-dioxetane, disodium salt) is the substrate most commonly used with alkaline phosphatase. It provides a sensitive and versatile substrate for alkaline phosphatase labels. Cleavage of the phosphate group produces an unstable anion that decomposes with the emission of light. The detection limit for alkaline phosphatase is 1 zmol (10^{-21} mol, 602 molecules).

As for other enzyme-based tests, the most popular format for rapid tests with chemiluminescent detection are rapid ELISA and lateral-flow immunoassay. Cho et al. [15] developed an immunochromatographic assay combined with horseradish peroxidase tracer that produces a light signal measurable on a simple detector. Cross-flow chromatography, a method previously developed for colorimetric test, was developed for cardiac troponin I.

A chemiluminescent signal was produced by adding a luminogenic substrate to the tracer enzyme complexed with the analyte on the chip. The luminescent signal was detected in a dark chamber mounted with a cooled charge-coupled device and the signal was converted to optical density for quantification. This system was capable of detecting cardiac troponin I present in serum at concentrations 30 times lower than those measured using the conventional rapid test kit with colloidal gold as the tracer [15].

It is difficult to compare sensitivity because of (for objective reason) there is no systematic comparison of different labels for the similar test systems with the same immunoreagents.

For cardiac troponin I, detection sensitivity was reported 0.027 ng/mL for chemiluminescent, 0.1 ng/mL for colorimetric and 0.82 ng/mL for the conventional kit with gold nanoparticles [15,22].

To make lateral-flow immunoassay with chemiluminescent detection one-step dry-chemistry test the chemiluminescent substrate concept relies on a delayed-release effect of chemiluminescence substrates (luminol enhancer and hydrogen peroxide generator) by an asymmetric polysulphone membrane. When the membrane was placed between the nitrocellulose running membrane and the substrate pad, substrates encapsulated in the substrate pad were released after about 5 min. In a model study, implementation of the sensor was validated by measuring the C-reactive protein level in human serum [24].

Quantitative chemiluminescent−based lateral-flow immunoassay was developed for the detection of 2,4,6-trinitrotoluene in real samples. To evaluate outcomes, a portable imaging device for chemiluminescent signal measurement based on a thermoelectrically cooled CCD camera was used; the analysis could be performed directly on-field. A limit of detection of 0.2 μg/mL 2,4,6-trinitrotoluene was obtained, which is five times lower than with colloidal gold−based lateral-flow immunoassay, developed employing the same immunoreagents [25].

In 2015 the smartphone camera was used as light detector, for image acquisition and data handling via a specific application in combination with a holder for strip cartridge. This provides a mini-dark box and an aligned optical interface between the camera and the lateral-flow immunoassay membrane for acquiring chemiluminescent signals (Fig. 1 from Ref. [26]). The method provided cortisol detection limit of 0.3 ng/mL [26].

3. DYES AND DYE-BASED LABELS

Coloured or luminescence dyes could be used as direct labels, substituting for radioisotopes in competitive or sandwich assays. Different series of dye molecules with desirable properties were synthesized and are commercially available. A wide range of coupling chemistries has been described in the literature [11].

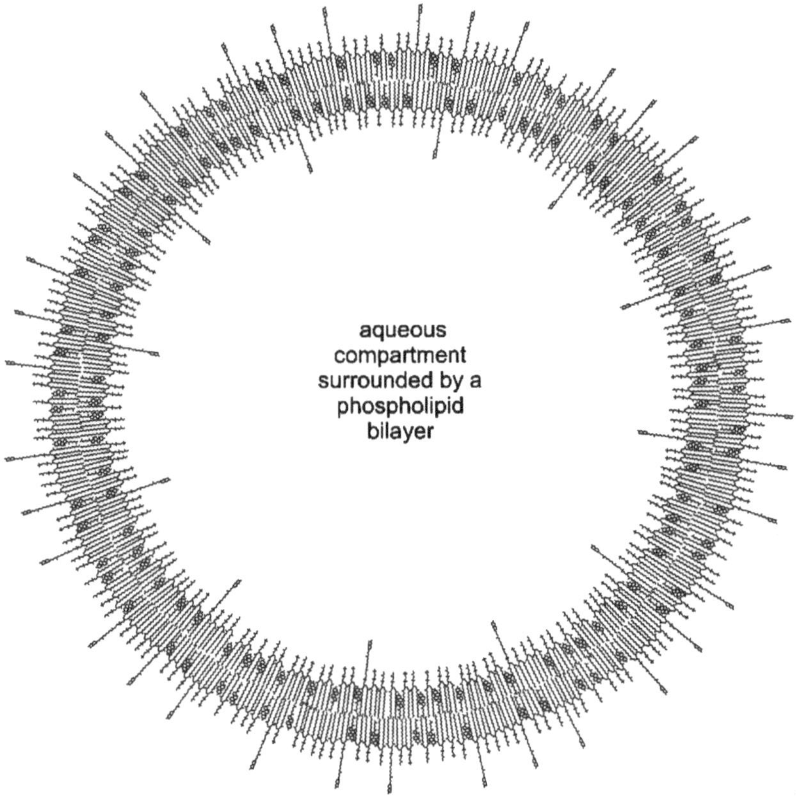

FIGURE 1 A liposome. Cross-section of a unilamellar vesicle with an aqueous interior surrounded by a bilayer membrane. In this example the liposome membrane consists of phosphatidylglycerol, phosphatidylcholine, cholesterol and biotinylated phosphatidylethanolamine [55].

3.1 Coloured Organic Molecules and Dye-Based Labels

Single dye molecules have not enough chromophore colour intensity, so 'multiloaded' labels were developed. One such label is the so-called 'colloidal dyes'

Only a few applications of colloidal dyes as labels for clinical assay were developed. Blue colloidal dye (D-1) and disperse dye (Dadisperse navy blue SP) were used as labels for immunochromatographic strips [27–29], red colloidal dye (R-3) was used for flow-through assay and the results were similar to those detected by routine ELISA [30]. It was shown that these kinds of labels are suitable for both quantitative and qualitative detection.

Coloured water-insoluble dye particles–streptavidin complex was used for the detection of pesticides (simazine and 2,4-dichlorophenoxyacetic acid) by noninstrumental competitive dot-immunoassay. The detection limit (4 ng/mL) was lower than the similar assay with horseradish peroxidase label (16 ng/mL), and the traditional competitive ELISA (12–16 ng/mL) were compared [31].

A colloidal dye dipstick test for multiple antigen detection combines sandwich ELISA and dot blotting. Coloured dye particles from a family of commercially available textile dyes of different colours served as detecting reagents. Colloidal dye and enzyme-labelled dot blot assays showed similar detection limits down to a sensitivity of 10 ng/mL and excellent agreement [32].

Another approach to enhance the chromophore colour intensity of dye molecules is loading them to polylysine of different molecular weights. The optimized dye chromophore using polylysine with a molecular weight of 189.4 kDa and a molar ratio (mol dye/mol amine group in polylysine) of 1.5 was used for labelling a model antibody in LFIA. The loading of polylysine with reactive dyes of different colours allowed [33] multiple analyte detection in a single qualitative or quantitative (by using the tabletop densitometer) immunochromatographic assay.

A new principle of label creation is by application of nanoparticles containing a precursor of highly coloured compounds. This type of label, which is based on organic nanoparticles, is loaded with the colourless indigo precursor 5-bromo-4-chloro-3-indolyl acetate and coupled with antibody for LFIA detection of a model protein as described by [34]. Through hydrolysis, this precursor produces a blue-coloured compound, 5-bromo-4-chloro-3-hydroxyindole, which forms an insoluble blue-coloured precipitate after oxidation, 5,5′-dibromo-4,4′-dichloro-indigo. For this purpose, a developing reagent composed of 2-propanol, NaOH and H_2O_2 was added to transform the precursor into the blue precipitate. Comparison of the proposed label with traditional AuNP labelling demonstrated better assay performance characteristics of the former (two times higher signal-to-noise ratio).

Another type of labels multiloaded with dye molecules is liposomes. Multiple molecules of red dye marker, sulphorhodamine B were loaded into liposomes to be used as coloured labels [35–38]. It is interesting to mention that these coloured liposomes were used for detection without lysis. More details about multiloaded liposomes and their application is described in Section 4.4.

3.2 Fluorescent Dyes

The simplest kind of luminescence emitters are single-molecule fluorescent organic dyes. Historically, fluorophores such as fluorescein, rhodamines and cyanine dyes have been used as cell and tissue labels in fluorescence microscopy and cell biology. Now several series of fluorescent dyes with improved fluorescent characteristics (high quantum yield, relatively large Stokes shift and chemical and photostability) such as Alexa Fluor from Invitrogen and Molecular Probes, PromoFlor from PromoKine, DyLight Fluor from Dyomics, ATTO Dyes from ATTO-TEC, Hilyte Fluor from AnaSpec are available.

These dyes have a potential to be applied as fluorescent labels for rapid tests. It is important to mention that the application of fluorescent labels to membrane tests is complicated by high light scattering caused by the membrane support. There is also interference of internal fluorescence originating from proteins (antibody), probe components, analytes like polycyclic aromatic hydrocarbons and some mycotoxins. Weak points of fluorescent dyes, compared to the nanoparticles, are low photostability, high quenching by the environment and concentration quenching [7].

Kim et al. [39] examined several fluorescent dyes including fluorescein isothiocyanate, rhodamine, Texas Red, Alexa Fluor 488 and Alexa Fluor 647, and reported the last one as being more stable and giving a higher fluorescence signal than others. This dye was reported in a highly sensitive LFIA for the detection of prostate-specific antigen [40], C-reactive protein [41] and human serum albumin [42] in whole blood without interference from blood compounds. For POC testing, it is important to apply the test directly to whole blood, because the application of the test to plasma and serum requires blood pretreatment. A similar test was also developed for albumin detection in urine [43]. A good correlation between the results of these tests and the ones consisting of more complicated and time-consuming methods was shown. In the area of environmental assays, a fluorescent dye–based LFIA was developed for the quantification of microcystins in surface water [39]. Sensitivity and reproducibility of immobilized-antibody and immobilized-antigen systems were compared for these low-molecular weight analytes and the superiority of the last one was shown.

Application of homemade [39] or commercial (*i*-CHROMA™, BioditechMed, Korea) laser fluorescent scanners allowed to obtain quantitative results. Cibitest's device FLORIDA consists only of an excitation lamp, and reading can be performed by the naked eye. Embedded Systems Engineering (Germany) created a miniature confocal optical sensor that can be modified with a wide variety of fluorescent labels for quantitative results' evaluation [7].

The disadvantages of traditional fluorescent chromophores are that they tend to have poor photostability, brightness and a short Stokes shift. This causes significant problems when working with samples with a naturally high background fluorescence. In addition, typically, only a limited number of fluorophores can be attached to a biomolecule before they begin to interfere significantly with binding specificity [44]. Consequently, targetting trace amounts of analytes may be almost impossible without signal amplification.

3.3 Fluorescent Dye–Based Labels

To increase the brightness of the labels, multiple fluorescent dyes can be coupled to a carrier and then to immunoreagents. This means a high label-to-antibody ratio and as a consequence, the signal intensity is larger than with standard conjugation of antibody to label. Another way to enhance the fluorescence intensity is by using particles (eg, polystyrene), loaded with organic fluorescent dyes. A variety of such microspheres is commercially available, such as FluoSpheres from Invitrogen. Similarly fluorescent dye-doped silica nanoparticles can be applied [45,46]. These approaches allow making labels brighter but do not solve problems with broad emission and small Stokes shift thus resulting in cross-talk between excitation and emission signals. Rhodamine B sulphonylchloride and other dyes can be covalently bound to and contained in silica nanoparticles of discrete spherical shape (30–80 nm). Compared to molecular, organic markers these fluorophore hybrid silica particles exhibit superior photostability and detection sensitivity. An immunoassay method for detecting trace level (down to 0.1 ng/mL) Hepatitis B surface antigen was successfully developed on this basis [47]. Synthesis approaches and analytical application of silica nanoparticles multiloaded with different reporters, including fluorescent dues was systematized by Knopp et al. [44].

High-sensitivity fluorescent (HSF) systems have been described (Sirigen Ltd) comprising polymer chains which act as ‘molecular antennae’ capable of collecting and transferring increased energy to standard fluorescent labels. It has been claimed that this technology increases signal by one to two orders of magnitude relative to standard fluorescence without increasing background [48].

Another type of enhanced labels is organic dye crystals. Solubility in aqueous environment and colloidal stability could be circumvented [49] by milling ($d \sim 500$ nm) and suspending a fluorogenic hydrophobic precursor fluorescein diacetate in sodium dodecylsulphate solution. Negative surface charge is introduced by the surfactant, rendering the particles colloidally stable and minimizing leakage of fluorescent molecules into surrounding water. In the next step, using the layer-by-layer technique, alternatingly charged polymer layers were adsorbed. Microelectrophoresis experiments revealed alternating negative and positive zeta-potentials with deposition of each successive polyelectrolyte layer, indicating that the alternate electrostatic adsorption of polyelectrolytes of opposite charge was successfully achieved. Protein immobilization onto the polyelectrolyte multilayer–coated particles was verified by the different surface properties of the microparticles with respect to surface charge under pH conditions above and below the isoelectric point of the proteins [49].

A high molar ratio of fluorescent molecules present in the microcrystal core to biomolecules on the particle surface was achieved. The applicability of the microcrystal-based label system was demonstrated in a model sandwich immunoassay for mouse immunoglobulin G detection. Following the immunoreaction, the fluorescein diacetate core was dissolved by exposure to a 1:1 mixture of dimethyl sulphoxide (to dissolve the particles) and 1N NaOH (to convert fluorescein diacetate into highly fluorescent, dianionic fluorescein), leading to the release of the fluorescent dye molecules into the surrounding medium. Amplification rates of 70- to 2000-fold (expressed as an increase in assay sensitivity) of the microcrystal label–based assay compared with the corresponding immunoassay performed with direct fluorescently labelled antibodies are reported [50]. Depending on the particle diameter, fluorescent dye-to-protein ratios of up to $\sim 10^8$ per affinity reaction are feasible [5].

Instead of sodium dodecylsulphate and multilayers, the adsorption of only one layer of an amphiphilic polymeric detergent, for example an alkylated poly(ethylene imine), is sufficient to stabilize the system and to provide an interface for the antibody attachment [51,52]. Depending on particle size and analyte concentration, the fluorescence signal was amplified up to 183,000-fold in comparison to a reference assay using traditional, molecular fluorescein isothiocyanate labelling. Limits of detection, however, were in the same order of magnitude throughout [52].

Preparation strategies and biological application of fluorescent organic nanoparticles are summarized by Suzanne Fery-Forgues [53].

4. LIPOSOMES

Liposomes are vesicles formed by a lipid bilayer with the hydrophobic chains of the lipids directed towards each other and the polar head groups of the lipids oriented towards the extravesicular solution and the inner cavity. Liposomes, first time described by Bangham in 1965 [54], are spherical vesicles composed of one or more phospholipid bilayers (uni-, oligo- and multilamellar vesicles) surrounding an aqueous cavity (Fig. 1 from Ref. [55]). The lipid molecules consist of hydrophilic head groups and hydrophobic tails; in aqueous solutions they self-organize to increase their solubility in the surrounding medium and minimize the surface-to-volume ratio. The size of the liposomes ranges from 50 to 800 nm.

Liposomes can be made easily by a variety of techniques from relatively inexpensive materials in relatively uniform size; their surface composition can be readily controlled by mixing phospholipids of different types, and the loading of markers could be performed during liposome preparation. In the film method [54], where a mixture of phospholipids and/or other amphiphilics is dissolved in an organic solvent, the solvent is removed at reduced pressure and the dry film of lipids is hydrated by adding a buffer with a water-soluble marker. In the reverse-phase evaporation method, an aqueous solution of markers is added to a mixture of lipids dissolved in an organic solvent and mixed till emulsion formation [56,57]. After removal of the organic solvent, dispersion of large liposomes is formed. In the detergent-dialysis method, where an aqueous solution of marker is added to a mixture of lipids in an organic solvent, a lipid film is formed when the organic solvent is removed, solubilizing the lipids by an appropriate detergent, resulting in lipid/detergent micelle formation, spontaneously producing liposomes when the detergent is removed by dialysis [55,58,59]. Markers such as fluorophores, enzymes or nanoparticles, present in the aqueous phase during the preparation, can be encapsulated in the liposome.

A variety of phospholipids with different polar head groups available for conjugation or reducing liposome aggregation, and hydrophobic regions possessing different chain lengths and saturation, are used to modify the properties of the resulting liposomes. The uniform, oriented structure of the lipid bilayer makes liposome surface properties predictable, and this allows control of the surface charge, the number of reactive groups and the hydrophilicity of the surface [60]. Different chemically active groups (including active groups for coupling biomolecules) can easily be incorporated on the liposome's surface in this way. It is important to mention that the properties and the amount of such groups can be varied over a wide range. This enables to control properties such as the size of the liposomes and the surface density of covalently bound biomolecules. Geometry and type of liposomes strongly depend on the preparation process; that is, the chemical composition and phospholipid concentration, mixing and centrifugation speed, injection procedure and type of the organic phase [61–63]. Liposomes can maintain their stability either in dehydrated form or in liquid solution for extended time periods. They were originally developed to study cell membranes and because of their ability to carry various water-soluble agents in their aqueous cavity, liposomes have been used in cosmetics and food industries. Later, liposomes found their application in drug delivery and clinical diagnostics [64]. Last decades, most research into applications of liposomes is in the field of drug delivery [65]. Clinical applications include the use of liposomes in gene therapy, as a carrier for genes [66].

The structure of liposomes offers the possibility of loading multiple signal-generating molecules. Molecules can associate with liposomes in several ways, including encapsulation within the aqueous inner cavity, partitioning within the lipid tails of the bilayer and covalent and electrostatic interactions with the polar head groups of the lipids [67]. A wide variety of molecules and nanoparticles can be encapsulated within liposomes: enzymes, visual/coloured and fluorescent dyes, QDs, electrochemical and chemiluminescent markers. Liposomes are promising for the development of sensitive immunochemical methods because they can carry a large number of receptor molecules and can be loaded with appreciable amount of markers due to their large surface area and internal volume [68]. A large number of markers have been encapsulated in liposomes, including organic molecules (dyes and fluorophores), enzymes, enzyme substrates, (small) proteins, RNA, DNA, ions, nanoparticles and radioactive isotopes [55].

First applications of liposomes to immunodiagnostics have been based on homogeneous assays which are generally turbidimetric or lytic. For liposomes loaded with luminescent markers or enzymes, immunochemical techniques were based on the lysis of liposomes and release of previously loaded markers. In lytic immunoassays, vesicles prepared with antigen on the surface release molecules encapsulated inside the aqueous core of the vesicle [69–71]. Other homogeneous assays use vesicles to enhance the measured turbidity resulting from the aggregation of antigen-coated latex beads [60,71].

Liposomes were applied in different heterogeneous assay formats, such as liposome immunosorbent assays [55,72,73], flow-injection immunoanalysis [74], lateral-flow assay [37], chemiluminescence biosensor [75], aptamer sandwich assays with fluorescent detection [76], multiplexed assays in microtitre plates [60,61] or microarrays [77].

Comparison of the signal of biotin enzyme–tagged liposomes to a biotin-tagged enzyme showed up to 100–1000 times higher signal with the liposomes for different immunosystems and different ways of signal generation [67]; a 500-fold lower detection limit was found measuring the fluorescence of lysed dye-encapsulated liposomes versus a single fluorescein-labelled probe [36]. Liposome advantages such as long-term stability of the encapsulated signalling molecules and the ease of labelling through the direct incorporation of hydrophobically modified nucleic acid probes into their lipid bilayers should be mentioned [36].

However, in most cases, the enhancement was not as significant as would be expected because of steric hindrance and multivalency. Liposomes have many biorecognition elements on their surfaces, thus one liposome can theoretically bind to several targets. Application of liposomes for analysis was reviewed by Edwards and Baeumner [67].

Lysing of the liposomes after the completion of the immunochemical reaction is where liposomes can be differentiated from most of the other multiloading carriers. In the case of lysing, the signal amplification is a result of liposome destruction with detergents, for example, a drawback of liposomes is their relative instability, although commercial liposomes are stable for up to 36 months. It is also difficult to dry and then reconstruct the liposomes without loss of their properties. This makes them hard to adapt to dry chemistry assays, such as LFIA. However, some examples of lateral-flow tests were described.

For application in lateral-flow tests, they could be targetted with DNA [67], RNA [78,79] and antigens [37,80,81].

Liposomes were used for multianalyte detection by multispots format for cross-reacting analytes [79]. Lateral-flow tests were mainly developed towards high-molecular weight analytes, but some examples were described for low-molecular weight molecules, such as aflatoxin B1 [37]. Variation of dyes could provide different coloured liposome labels: methyl blue gives the liposomes a blue colour [81] and sulphorhodamine B makes liposomes pink [36], and the latter could also be used for fluorescent measurement [79,80]. Aequorin is a photoprotein isolated from luminescent jellyfish in liposomes and it was also applied as a luminescent label [82].

For visual dyes, the colour intensity may be measured semiquantitatively by visual examination. However application of readers (eg, QuadScan reflectance photometer (KGW Enterprises, Inc., Elkhart, IN)) can provide more accurate quantitative results.

4.1 Markers and Encapsulation Strategies

Assays based on the measurement of dye fluorescence are inherently more sensitive than the measurement of dye UV absorption [60]. Fluorescent markers, initially employed as a means to study membrane properties, are most commonly used because these markers can be measured rather easily and are very sensitive. Carboxyfluoresceins [83,84], sulphorhodamine B, Cy3 and calcein [85] are used as markers.

Apart from these small molecules, larger markers can also be entrapped, including enzymes and nanoparticles. Enzyme examples are alkaline phosphatase, horseradish peroxidase and glucose oxidase and cofactor example is flavin adenine dinucleotide. Larger molecules tend to leak less through the liposome membrane than smaller molecules [55]. If an enzymatic marker is used, additional incubation period is necessary for enzymatic conversion of the substrate, resulting in an instantaneous amplification. Nanoparticles, such as QDs also could be loaded in liposomes.

If the marker is trapped in the internal volume of the liposome's aqueous core, the concentration of the encapsulated content should be as high as possible, leakage should be minimal and if lysis is needed for results' evaluation, the contents should be released only on the target site. The encapsulation of dyes and other small marker molecules inside a vesicle can be achieved readily by suspending the phospholipid in an aqueous phase containing the marker. Dye concentrations used in these preparations are in the order of 0.01−1.0 M [60]. For example, liposomes with an external diameter of 57 nm and a bilayer thickness of 4 nm [86] have an internal volume of 6.2×10^{-20} L. If a 0.1 M solution of dye is encapsulated in this space, in principle it would lead to about 3700 entrapped dye molecules per vesicle [60].

The advantage of utilizing liposomes with entrapped fluorescent molecules is that the fluorescence is self-quenched until disruption by detergent, which allows an extra degree of control.

While encapsulation in the internal aqueous core is the most popular way to prepare loaded liposomes, the encapsulated content could diffuse out of the liposomes during the storage, and with higher concentrations of encapsulated material the diffusion rate can be faster.

To solve this problem several approaches could be used:

- Making encapsulated content an integral component of the membrane structure by attaching directly to the lipids forming the bilayer.
- Minimization of encapsulated content leaking by modification of bilayer properties or covering with additional layer.
- Using of hydrophobic encapsulated content.

The advantage of direct attachment of labels to the liposome surface is more apparent for enzymes. In these cases it is not possible to achieve high concentrations of enzyme solutions for encapsulation, hence it would be typical to have only a handful of entrapped enzymes per liposome.

Heath et al. [87] and Jones et al. [60] described the covalent attachment of horseradish peroxidase to the outer surface of liposomes. This development offers the opportunity for the enhancement of ELISA signals and sensitivity using ligand- and enzyme-conjugated vesicles [60]. In some instances it may be possible to attach at least as many labels to the surface of a liposome as can be entrapped within the aqueous core. A large number of enzyme molecules (>100) can be attached to each liposome surface, while at the same time maintaining good specificity and binding strength to the plates and in optimized conditions the signals generated by the bifunctional horseradish peroxidase and biotin-conjugated liposomes were found to be about 100 times greater than those generated by biotinylated horseradish peroxidase [60]. At the same time it was found, after conjugation with liposome surface the enzyme specific activity decreased between 35% and 15% when compared to native enzyme. In [88] it was shown that about 7800 fluorescent molecules could be attached to a 57-nm diameter liposome.

Liposomes could be sensitive towards external influences in experimental conditions and during the storage [89,90]. Semipermeability of liposome membrane, so essential for medical and pharmaceutical aims, could do a bad turn in chemical analysis. The possible osmosis of encapsulated content through the phospholipid bilayer would result in analytical signal decrease, which is undesirable for immune-label. Leakage of the phospholipid membrane can be halted through the coverage of liposomes with a polymer net [91,92] or silica cover [93,94]. Silica coverage prevents the aggregation of liposomes and the leakage of their content, increasing their stability and facilitating their desiccation and therefore their storage [95].

Because of amphiphilic nature of phospholipids, liposomes are able to encapsulate hydrophilic compounds (such as water-soluble QD) within a liposome's internal volume and hydrophobic substances (such as water-insoluble QD) within the liposome's membrane. With both liposomes loaded with hydrophilic and hydrophobic QDs, the luminescent analytical signal was detected without lysis directly after addition of the labelled conjugate. However, it was found that in the case of liposomes loaded with water-soluble QDs, luminescence intensity increased after liposome lysis. It is important to mention that lysis did not extend the analytical procedure because the buffer, which was used to redissolve the content of each well in the last procedure step, contained the surfactant (in this case, 0.1% Tween 20) [95].

4.2 Formats of Liposome Immunoassay

A variety of markers loaded into liposomes in combination with different variants of immunoassay predicts wide possibilities of application in clinical assay and also in food and environment control. The main advantage is the application of the immunoassay format and detection techniques and equipment of the well-known ELISA and LFIA procedures.

Heterogeneous formats. As for all heterogeneous immunoassays, liposome immunosorbent assays (LISA) always use separation of the specifically bound liposomes from the free liposomes. LISA can be regarded as modifications of 'normal' ELISA. The substitution of the enzyme by a liposome gives the assay the benefit of a large amount of immediately available marker. Lysis of the liposomes can be performed with a detergent, such as Triton X-100 or *n*-octyl-b-D-glucopyranoside. Both substances effectively and quickly lyse all liposomes.

Competitive LISA with antigen conjugated liposomes. As for similar ELISA format, a mixture of antigen and antigen-coated liposomes is added to antibody-coated solid phase and is followed with a washing step to remove unbound liposomes. The LISA different is an addition of detergent, which lyses the bound liposomes, and finally either the marker is determined, or in the case of enzymatic marker a substrate is added and the formed product is measured. If enzyme is not included into aqueous core, but attached to the lipid bilayer, the detergent step can be omitted [96].

Noncompetitive sandwich LISA. An antigen is added to an antibody-coated solid-phase followed with captured antibody–coated liposomes [60,96,97]. As a last step detergent is added and the marker is released. Vonk and Wagner [97] used time-resolved fluorescence as a detection method, with a europium–diethylenetriaminepentaacetic acid complex encapsulated in liposomes.

Homogeneous immunoassays. Common to all homogeneous assays is the fact that the components of the assay are mixed in a single vessel and there are no separation steps. So the specific release of the marker is also measured in that vessel. Particular feature of homogeneous liposome immunoassays consists of the detection of a natural lytic agent, such as complement or mellitin. The biological function of both agents is to destroy foreign cells, the former as part of normal defence mechanisms and the latter as a component of bee or snake venom [55]. Thus, natural properties of lytic agents are used for their detection. The most frequently used markers are the fluorophore carboxyfluoresceine and the alkaline phosphatase enzyme AP. A drawback could be the slow and sometimes incomplete lysis.

Complement is a group of serum proteins [98], which in their natural environment cause destruction of invading foreign cells [99]. When foreign cells enter the body they are opsonized by antibodies, which attach themselves to the cell membrane surface. Complement proteins bind to these antibodies in a specific order after which the target cells are lysed [55,100]. Because liposomes resemble cells very closely, complement can be used to lyse antibody-associated liposomes, as was first demonstrated by Kinsky in 1972 [101]. This way the presence of a lytic agent in a biological sample could interfere with a liposome immunoassay and this possible complication should be kept in mind for experiments with liposome planning. If the analyte in the sample thermostable, complement can be inactivated by heating the sample for 30 min at 56°C.

Complement lysis is dependent on the structure of the liposome and can be relatively fast (for large unilamellar liposomes) or rather slow (for multilamellar liposomes), but in both cases considerably slower and less efficient than detergent lysis [102,103]. Complement-mediated assays comprise the largest group of liposome immunoassays; the range of antigens is not limited in size.

The general formats of all complement-dependent assays are very similar, and are illustrated in Fig. 2 (from Ref. [55]). In a direct competitive format, antigen-coated liposomes are mixed with antigen and antibody incubated mixture (Fig. 2A). Complement addition causes immunospecific lysis of antibody-bound liposomes. The detection method depends on the properties of the encapsulated marker.

In noncompetitive, sandwich type assays, antibody-conjugated liposomes are incubated with antigen and a second unconjugated antibody, which is directed against a second epitope on the antigen (Fig. 2B) [104,105]. Formation of sandwich immunocomplex joining complement and liposomes results in liposomes lysis.

Mellitin, a natural protein from bee venom, like complement, is able to lyse the lipid bilayer membrane of liposomes completely, although rather slowly [106]. Mellitin was introduced for liposome immunoassays by [107]. These immunoassays use a mellitin-antigen conjugate. Its action depends on the inhibition of the lytic effect through the binding of an antibody Fig 3 (from Ref. [55]). For mellitin-based assays, where the lysis by mellitin is inhibited by antibody binding, the range of antigens is limited for small-sized analytes.

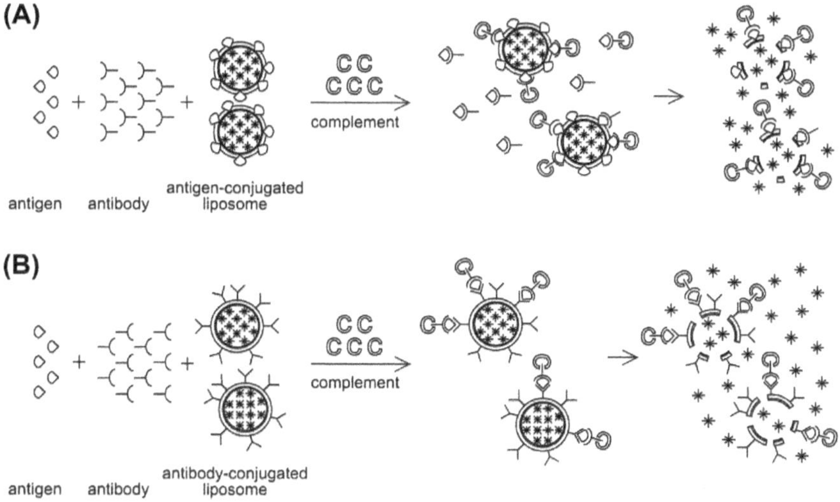

FIGURE 2 Homogeneous complement liposome immunoassay formats: (A), a competitive and (B), a noncompetitive format [55]. *This picture is from H.A.H. Rongen, A. Bult, W.P. Van Bennekom, Liposomes and immunoassays J. Immunol. Methods., 204:105–133, 1997.*

This kind of assay can also be performed using a microtitre plate as solid phase.

Typical example of complement-based immunoassay is low molecular–weight trichothecene mycotoxin T-2 detection in a homogeneous competition inhibition assay based on complement-mediated lysis of liposomes. The T-2 mycotoxin was incorporated into unilamellar liposomes. Carboxyfluorescein, was entrapped in the liposomes as a release marker, where it was self-quenched because of high local concentration. A monoclonal antibody specific for T-2 mycotoxin and a polyclonal anti-mouse secondary antibody since the anti-T-2 IgG1 does not activate complement were used. In the absence of free T-2, the liposomes were lysed within 30 min after the addition of complement, releasing carboxyfluorescein into the surrounding buffer. In the presence of

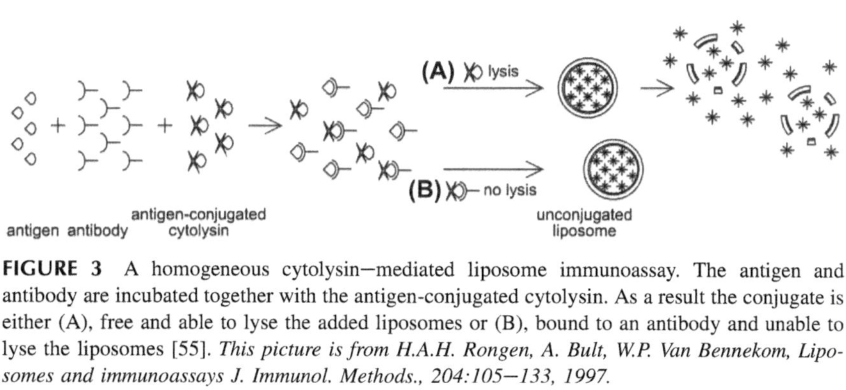

FIGURE 3 A homogeneous cytolysin–mediated liposome immunoassay. The antigen and antibody are incubated together with the antigen-conjugated cytolysin. As a result the conjugate is either (A), free and able to lyse the added liposomes or (B), bound to an antibody and unable to lyse the liposomes [55]. *This picture is from H.A.H. Rongen, A. Bult, W.P. Van Bennekom, Liposomes and immunoassays J. Immunol. Methods., 204:105–133, 1997.*

free T-2 toxin, the binding of antibodies to the liposomes was reduced, causing a corresponding decrease in lysis. This assay proved to be sensitive to T-2 toxin levels as low as 2 ng, which is 10-fold more sensitive than the present enzyme immunoassay using the same antibodies [108]. Similar assay was developed for low molecular–weight fluoroquinolone antibiotic enrofloxacin in carp and chicken muscle [109].

5. COLOURED NANOPARTICLES

Coloured reporter systems are the most popular labels for rapid tests. Sensitivity up to the low picomolar range was reported for such labels, even by visual inspection. Unlike the previously described kinds of labels, common properties of nanoparticles are photostability, long shelf life and size-dependent mobility. Usually, labels with sizes of 15–800 nm allow an unobstructed flow through the LFIA membranes [2]. Commonly, conjugation of nanoparticles with biomolecules is realized by surface modification with amino or carboxylic groups.

5.1 Colloidal Gold

Colloidal gold is the most popular label for rapid tests and it is also often referred to as the most stable of all colloids. The first information on colloidal gold can be found in tracts by Chinese, Arabic and Indian scientists, who obtained colloidal gold as early as in the fifth and fourth centuries BC and used it, in particular, for medical purposes (the Chinese 'gold solution' and the Indian 'liquid gold') [110]. Despite the centuries-old history, a 'revolution in immunochemistry', [111] associated with the use of gold particles in biological research, took place in 1971, when W.P. Faulk and G. M. Taylor published antibody conjugation with colloidal gold nanoparticles (AuNPs) for direct electron microscopic visualization of *Salmonella* surface antigens, representing the first time that a colloidal gold conjugate served as an immunochemical marker [110,112].

AuNPs have unique physical properties which are dependent on their size, shape and the interparticle distance. This simplifies biological application of this label by removing the necessity of covering the toxic core with inert material. The functionalization of AuNPs could be achieved by linking various biofunctional groups such as amphiphilic polymers, silanols, sugars, nucleic acids and proteins, via the strong affinity of the gold surface with thiol ligands. AuNPs are low susceptible to aggregation during the preparation of the test devices. Preparation, properties and analytical and biomedical application of AuNPs was reviewed by Khlebtsov group [110,113,114], Wang and Ma [115] and Chun P [116].

The colours of AuNPs, the basis for their application as markers, are due to the presence of a surface plasmon resonance (or localized surface plasmon resonance) absorption band. This absorption band results when the incident photon frequency is in resonance with the collective excitation of the conductive electrons of the particle [110]. The colour of AuNPs and their correspondent wavelength of maximal absorbance strongly depend on the size and shape of the gold nanoparticles. The brightness of AuNPs depends on their size which can be controlled easily during manufacturing. The sensitivity of 30-nm particles is 2–4 times better compared to 15-nm particles [117]. Bao et al. [118] found that 80-nm gold nanospheres are much more sensitive than the standard fluorescent labels (Cy-3, Cy-5). However, according to Khlebtsov and Khlebtsov [119], the experimental work with particles exceeding 30–40 nm is difficult because of colloidal stability problems.

Colloidal gold conjugation has become a standard method in immunochromatographic tests in very different areas: POC, then food control and recently environmental control. One of the most popular AuNP–based formats is LFIA. It has wide application in the areas of POC, but less for food control and environmental purposes. Spreading of the colloidal gold–based LFIA is hampered by the low sensitivity of the label. Nowadays an attempt to use AuNPs-based LFIA for lead ions detection in water samples has been published [120]. Also DNA- and aptamer-based lateral-flow dipsticks with AuNP labels have been reported for detection of DNA [120], adenosine [121] and lead [122]. A variety of readers for AuNP-based LFIA is commercially available. Application of AuNP in multiassay is only possible by using separate test zones.

By using of a galvanic replacement reaction between the Ag atoms of silver nanocubes and Au ions of tetrachloroauric acid particles of different colours were obtained. Depending on the Ag/Au conversion ratio, the particle plasmon resonance was tuned from 450 to 700 nm and the suspension colour changed from yellow to blue. The multiplex capability of yellow, red and blue particles was illustrated in a dot immunoassay format [123].

Several approaches for signal enhancement have been developed and two main ones can be marked out in this area (Fig. 4). The first one is modification of AuNPs without complicating the further assay procedure. In the second approach, assay sensitivity is improved by incorporating additional steps to the procedure; LFIA became a multiple-step format in this way.

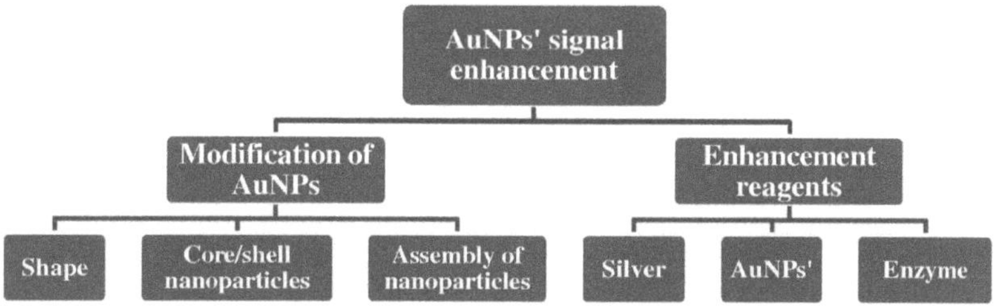

FIGURE 4 Strategies for gold nanoparticles signal enhancement.

5.1.1 Modification of AuNPs

Modification of AuNPs allows increasing sensitivity without additional assay steps. Core-shell nanoparticles are becoming more popular because of their brightness. Khlebtsov and Khlebtsov [119] performed theoretical estimations and showed that the dot extinction of 100 nm silica–gold nanoshells can be 1000 times higher than that for the same number of 15 nm AuNPs.

Multibranched blue gold nanoflowers (diameter about 75 nm) were used to enhance the detection sensitivity of LFIA because of their high optical extinction from the tips and the core–tip interactions required for electromagnetic field enhancements. Compared with spherical AuNPs, nanoflowers exhibit large surface-to-volume ratio, which can improve the antibody amounts of immobilization. The detectable concentration of an analyte (aflatoxin B1 in rice) was 10 times lower than that of traditional AuNPs-based LFIA [124]. Liao and Li [125] published an immunodipstick with core–shell silver–gold nanoparticles which show higher intensities compared to pure AuNPs. The authors also mentioned the application of core–shell nanocomposites resulting in a strong enhancement of sensitivity of aflatoxin B1 detection while possessing a reproducibility and stability comparable to traditional AuNPs.

Application of a detector reagent consisting of nano-Fe_2O_3 particles as core and AuNPs as a shell for LFIA detection of aflatoxin B2 allowed Tang et al. [126] to increase sensitivity threefold compared to conventional AuNPs. Xu et al. [127] described a sandwich LFIA for detection of proteins by using AuNPs-decorated silica nanorods as the label with an LOD of 0.01 ng/mL, which is 50 times lower than that of traditional GNP–based LFIA.

5.1.2 Signal Enhancement Procedures

Signal enhancement step is usually realized after the standard assay procedure. Two different detection strategies can be used this way: the first one involves just monitoring the red colour of the AuNPs and (if the sensitivity is not enough) the second one involves monitoring of signal after an enhancement procedure [16].

5.1.2.1 Silver and Gold Enhancement

Gold or silver, which can be nucleated on the AuNPs surface, has been widely used for colorimetric signal amplification. Gold or silver intensification is chemically based on the fact that gold or silver ions are reduced and deposited on the AuNP surface [128].

Upon addition of silver-containing enhancement buffer, metallic silver deposits onto gold nanoparticles, causing an obscuration that can be optically measured by a CCD camera or scanner or which can be visually detected. The chemical basis for silver intensification via colloidal gold, known as autometallography, dates back to 1930 and its application as gold tracer linked to an antigen–antibody reaction was first demonstrated by Scopsi et al. [129]. Gold provides a route for the transfer of electrons from the reducing agent in solution to silver ions bound to the gold surface. This results in specific deposition of metallic silver at the site of gold labelling, and the silver can then participate again in the catalytic reaction [130]. The sensitivity of immunogold detection can be increased about 100-fold by silver enhancement [131]. This approach is restricted by the need for additional steps as well as the strong requirements for preceding washing steps for removing interfering ions such as chlorine.

Some examples of silver enhancement applications for different formats have been published. A multiwell chip was developed for human immunoglobulin G detection using a CCD camera [132]. Flatbed scanning was used for detecting a model antigen [133]. Wu et al. [134] used polydimethylsiloxane–AuNPs composite film as basis together with silver enhancement for colorimetric detection of cardiac troponin I. The obscuration due to silver deposition was relative to the amount of antigen because of the difference in inhibiting-ability between silver enhancement blocked with bovine serum albumin film on the one hand and antibody–antigen complex on the other. Cross-flow chromatographic analysis was described in 2010 [130] as a variant of the LFIA procedure. It is interesting to mention that silver enhancement could be applied not only for AuNPs, but also to increase sensitivity of silver nanoparticle–based LFIA [133].

Similar to silver enhancement, gold enhancement can provide signal amplification by increasing the signal-to-background ratio of the detection zone by about 25%. Rohrman et al. [135] described LFIA for the sensitive detection of HIV-1 RNA by using gold enhancement reagents to enlarge AuNPs and provide intense colorimetric signals. Li et al. [136] developed an AuNPs-based LFIA for 153 detecting avian influenza and Newcastle disease viruses. The fabricated strip exhibited a 100-fold increase in sensitivity compared with commercial LFIA.

AuNPs' enhancement. In the group of 'sensitizers' application of AuNPs itself as an enhancement reagent could be singled out. In the AuNPs' application-group of 'sensitizers', it can be singled out itself. After performing the normal method, the AuNPs conjugated with primary antibody will be captured by secondary antibodies on the test line of the strip [137,138]. To keep an assay in one-step operation it is possible to place 'sensitizers' on the additional conjugate pad [139]. As a result, AuNPs accumulate at the test line and the sensitivity is higher. Nagatani et al. [137] reported a 50-fold decrease in limit of detection using this procedure (25 pg/mL and 1.0 ng/mL of human chorionic gonadotropin hormone with and without 'signal enhancement' procedure). A variant of this approach, however without any additional steps, also includes two types of bioconjugated AuNPs encapsulated in different pads. One type being AuNPs conjugated with antibody and blocked with BSA, and another being AuNPs conjugated with antiBSA antibody. Choi et al. [140] found that the size of the two conjugates is very critical for the detection limit.

When 10 nm AuNPs for the first and 40 nm for the second conjugate were used, the sensitivity increased about a 100-fold compared to the conventional LFIA. An additional advantage of the AuNP enhancement procedure is that the same reader can be applied as for normal AuNP–based method. This approach was first applied for macromolecule detection [139] and later was broadened for low-molecular weight analyte bisphenol A detection [140].

Enzyme enhancement. Application of AuNPs as a label enhanced with horseradish peroxidase (HRP) allows an improved optical detection. After chromogenic substrate application, a darker colour appeared with enhanced intensity compared to the red colour of the unmodified AuNPs. Parolo et al. [16] compared three different substrates and found increase in sensitivity up to one order of magnitude.

Use of the catalytic properties of AuNPs allows their application for chemiluminescence detection. AuNPs could trigger the reaction between luminol and $AgNO_3$, accompanied by light emission. This approach has been used for human immunoglobulin G detection [141].

A principally other way for AuNP assays with visual detection based on their distance-dependent properties was introduced for the first time by Elghanian et al. [142] for polynucleotides detection based on the AuNPs colour change. When individual AuNPs come into close proximity (the centre-to-centre distance is usually smaller than 2.5 times the diameter of the AuNP), the surface plasmons of individual AuNPs combine (so-called interparticle surface plasmon coupling), which results in a change of colour from red to purple and blue.

A similar principle was used in homogeneous immunoassay, using human serum albumin as a model analyte. Binding of antibodies to functionalized nanoparticles causes a shift of the visible absorption maximum of AuNPs. Quantification of the analyte could be obtained after competitive inhibition [143].

The distance-dependent properties were also used for rapid colorimetric detection of small analytes as for example melamine, dopamine and ascorbic acid. Because of the presence of three amino-groups, melamine could rapidly induce aggregation of label-free AuNPs, resulting in a red-to-blue (or purple) colour change. Li et al. [144] showed that the concentration of melamine in raw milk can be determined by monitoring with the naked eye or a UV–vis spectrometer with an LOD of 0.4 mg/L. The method is rather simple and the whole process takes only 12 min, including sample pretreatment (centrifugation with trichloroacetic acid, pH adjusting and filtering). Guo et al. [145] proposed a similar method for milk and infant formula with a detection limit of 1.0 and 4.2 ppm, respectively, by naked eye, and 0.15 ppm and 2.5 ppm, respectively, with UV–vis-spectroscopy within 30 min (sample preparation includes ultrasonification with trichloroacetic acid and chloroform, centrifugation and pH adjustment). Comparison of several modifiers showed that the modification with cysteamine, aimed at weakening the electrostatic repulsion force between the AuNPs, increased the signal intensity about a 100-fold compared to the unmodified AuNPs [146]. A similar approach was used for the detection of dopamine; the presence of Cu^{2+} ions increases sensitivity through complexation of two dopamine molecules [147]. Ascorbic acid could also induce a rapid aggregation of azide- and alkyne-functionalized AuNPs in the presence of Cu^{2+}, thereby resulting in a colour change [148].

This distance-dependent approach was also realized in a solid-based application, which is more suitable for practical use. Liu and co-workers [121] first developed a lateral-flow device where AuNPs aggregate upon addition of a specific target analyte after which they dissociate into well-dispersed AuNPs. These dispersed AuNPs are modified by a biotin flow along a cellulose-based membrane and are captured by a coated line of streptavidin on the membrane. A red colour appearing on the streptavidin line indicates the presence of the target analyte. Inspired by this work, Zhao et al. [149] have demonstrated the feasibility of using an AuNP-based dot-sensing platform on paper strips. DNA-cross-linked AuNP aggregates were spotted on paper, and a red colour signal was 'turned on' in the presence of the target DNA by redispersion of the AuNP aggregates on paper. New sensor formats, based on colorimetric detection are reviewed by De la Escosura-Muciz et al. [150].

5.2 Colloidal Carbon

The black colour of carbon nanoparticles can be visually detected with high sensitivity and was firstly introduced as a label for a rapid immunochemical test in 1993 by Van Amerongen et al. [151]. Preparation, functionalization and application of carbon nanoparticles for rapid immunotests were reviewed by Posthuma-Trumpie et al. in 1993 [151]. Carbon nanoparticles are very inexpensive, available in large batches and they are very suitable for quantification of results using 'gray pixel' processing because of their 'black-on-white' test results.

Due to the high extinction of carbon black, the detectable concentrations of the pure particles are 0.04 ng/mm^2 (0.02 amol/mm^2) by use of a scanner and 0.2 ng/mm^2 (0.1 amol/mm^2) by the naked eye [152]. These authors compared the detection ability to those obtained with enzymatic labels: alkaline phosphatase (AP) using a substrate yielding a chemiluminescent signal (0.02 amol AP/well), the use of β-galactosidase and a substrate resulted in a fluorescent signal (0.3 amol β-galactosidase/well) and HRP using a substrate rendering a coloured signal (5 amol HRP/microtitre well). The sensitivity of LFIA with carbon black labels was reported to be equal to that of the corresponding ELISA [153]. It was also shown that carbon black had a remarkably low detection limit of 0.01 μg/mL compared to 0.1, 1 and 1 mg/mL for silver-coated gold nanoparticles, gold nanoparticles and polystyrene beads, respectively [154,155].

Opposite results were obtained by Collin et al. [156]. A colloidal carbon—antibody conjugate was tested as a label for sporidesmin A LFIA, but its sensitivity was less (25 ng/mL) than for colloidal gold conjugates (up to 4 ng/mL). Carbon black labelling was applied in the development of LFIA for detection of human chorionic gonadotropin [157], immunoglobulin E [144], *Schistosoma* circulating cathodic antigen [154], Shiga toxin producing *Escherichia coli* virulence factors [158], fungal α-amylase [159], carbaryl [160] and thiabendazole [161] in fruit juices, phytoregulator forchlorfenuron in fruit samples [162] and also for simultaneous detection of *Listeria* spp. and *L. monocytogenes* [163].

High sensitivity compared to other LFIAs, and even 50 to 100 times more sensitive than corresponding enzyme-based immunoassays, has been achieved by utilizing large carbon black nanostrings (a size corresponding to a spherical particle of about 200 nm in diameter) in combination with sensitive image-scanner detection [164]. It is interesting to mention that the application of multiwalled carbon nanotubes as a label for LFIA allows quantitative electrical detection [165].

5.3 Colloidal Selenium

Rust-coloured colloidal selenium (λ max = 540 nm) was employed for detection of lipoprotein A in plasma. Lipoprotein A—coated colloidal selenium was stable from two months in liquid form up to over eight months in lyophilized form [166]. But since this first publication there were no further applications of this label.

5.4 Colloidal Iron Oxide

Fe_3O_4 nanoparticles are applied as labels mostly because of their magnetic properties. However, their optical properties could be helpful when used as coloured labels. The disadvantages of iron oxide particles in comparison with AuNPs are:

1. The absorption spectrum of iron oxide particles is wide, covering almost the whole visible range due to the intraband transition.
2. The dark brown colour of Fe_3O_4 nanoparticles is apparently not as bright as that of the AuNPs. Nevertheless, the integral molar absorption coefficient of Fe_3O_4 nanocrystals within the visible light range is rather comparable to that of the AuNPs. Furthermore, aggregation hardly changes the absorption properties of iron oxide nanoparticles, where it does change for AuNPs [167,168].

Fe_3O_4 nanoparticles were used as coloured labels for LFIA in the detection of the pesticide paraoxon methyl, obtaining a detection limit of 70 ng/mL. More interestingly, application of Fe_3O_4 nanoparticle aggregates demonstrated that the detection limit could be decreased over 40 times, reaching 1.7 ng/mL. Visual detection limits were 1000 and 150 ng/mL for nanoparticles and aggregates as labels, respectively. The Fe_3O_4 nanoparticle aggregates were prepared by cross-linking surface carbonyl groups containing Fe_3O_4 nanoparticles with poly-L-lysine [167].

An additional advantage of this label is the possibility to use two analytical signals (optical and/or magnetic) with one test without additional steps.

6. LUMINESCENT NANOPARTICLES

Reporter systems with luminescent detection are now widely used in research and clinical diagnostics because they provide high sensitivity and capability for simultaneous use of multiple labels with different spectral characteristics (multiplexing). In some cases the sensitivity of luminescent labels is comparable to enzymatic labels, but the assay procedure is essentially simpler. As the chemical nature of luminescent labels with similar optical properties can be very different, they are classified in this section according to their principle of luminescence emission.

In general, emitters can be categorized as down-converting and up-converting. Down-converting emitters absorb a photon and then emit a photon with lower energy. This is the more common process and it is exhibited by fluorescence, phosphorescence and most of the between-molecules energy-transfer processes. In contrast, up-converting phosphor absorbs two or more photons with an energy that is lower than the energy of emitted photon.

6.1 Quantum Dots

Another type of labels useful for visual and instrumental detection of rapid test outcomes are QDs, which are inorganic luminescent semiconductor nanocrystals, exhibiting size-dependent luminescence emission spectra. Since the 2000s QDs' applications in immunoassays can be mentioned as the most expanding area of immunoassay labelling. A myriad of QD-based assays were described in scientific literature over the past decade, but despite their unique properties and photophysical advantages, QDs are yet to find acceptance as standard fluorescent reagents in rapid tests. Application of QDs in immune- and bioassay was critically reviewed [170–173]. One of the most developing areas of QD application in assay is Förster fluorescence resonance energy transfer (FRET) [174] (Section 3.3), but till now it is difficult to classify FRET as a popular rapid test format.

Various types of QDs were synthesized for different research applications such as InP, InAs, GaAs, GaN, ZnS and ZnSe, as well as QDs with more heavy atoms such as CdTe, HgSe or PbSe. However, the most popular core for bioassay application is CdSe because different-sized CdSe dots emit light across the whole visible spectrum. Usually, the CdSe core radii vary between 1 and 6 nm, which is known to be smaller than the bulk Bohr excitation radius (for CdSe it equals to 6 nm). When the radius of the QDs is smaller than the bulk exciton Bohr radius, it strongly influences the energy bands under the quantum confinement. Combined with a decrease in QD radius, this causes a blue shift in emission. During fabrication, the diameter of the QDs can be selected to achieve emission in a variety of colours. However, uncovered CdSe nanocrystals do not have high quantum yields and are not stable enough in water solutions due to surface-related trap states acting as fast nonradiative de-exciting channels for photogenerated charge carriers [175,176]. Coating of core nanocrystallites with higher band gap inorganic materials (shell) has been shown to improve the photoluminescent quantum yields by passivating the surface nonradiative recombination sites. Several wide band gap semiconductors (ZnS, CdS or ZnSe) can be used as shell material. Zinc sulphide is the most popular shell material due to its wider band gap (energy difference between the valence band and the conduction band), thus allowing the efficient confinement of both the photogenerated electrons and holes in the nanocrystals' core. An additional middle shell (CdS or ZnSe) located between the CdSe core and the ZnS outer shell could be added to reduce the nanocrystals' internal strain caused by CdS and ZnSe intermediate lattice parameters compared to those of CdSe and ZnS [7].

QDs' extinction coefficients are $\sim 10^5-10^6\ M^{-1}\ cm^{-1}$, depending on the particle size and the excitation wavelength which is about an order of magnitude higher than organic dyes. The quantum yields of CdSe QDs vary up to 40%. In comparison to classical organic dyes such as rhodamine 6G and fluorescein, CdSe nanocrystals show lower quantum yields at room temperature, but the lower quantum yields are compensated by their larger absorption cross-sections. Chan and Nie [169] estimated that each CdSe/ZnS QD is about 20 times brighter and 100–200 times more stable than rhodamine 6G.

Opposite to organic fluorescent dyes, QDs have symmetric and narrow emission bands and their absorption spectrum reaches into the UV area regardless of their size. So, their spectral characteristics allow simultaneous excitation of particles with different sizes using a single wavelength resulting in emission at multiple wavelengths. These characteristics make them ideal labels for detecting multiple binding events in one spot [170,177]. The weak points of QDs, such as toxicity, water-insolubility and absence of functional groups available for bioconjugation were overcome by the introduction of QDs to silica nanoshells [178]. Another way to achieve water-solubility and biocompatibility is conjugation with a bi-functional molecule (eg, mercaptoacetic acid) [169]. QDs have been widely applied for biolabelling in the fields of cell biology, molecular biology, genomics and medical diagnosis, and are commercially available in free and conjugated forms (Invitrogen, Evident technologies). Water-soluble CdSe–ZnS nanoparticles, compatible with conjugation methods in aqueous conditions, are usually prepared using a stepwise procedure. This consists of core CdSe nanocrystal growth, ZnS coating, capping with an organic layer and finally size selection precipitation [179].

Broad excitation and narrow emission bands allow simplifying the use of QDs for multiassay and also minimize the matrix's influence. This latter aspect is especially important for blood sample analysis. Also, the luminescent lifetime of QDs tend to be in the range of 30−100 ns which is slightly longer than those of organic dyes (1−5 ns), but much shorter than those of lanthanide probes (1 μs−1 ms). These excited-state decay rates last much longer than the background fluorescence and Raman scattering of most sample matrices. One can therefore use time-gated detection to selectively reduce or remove background fluorescence [116]. Functionalization of QDs by different bioconjugation and capping approaches were described by Frasco and Chaniotakis [180]. Application of QDs for chemical and biological detection, diagnostics and imaging was reviewed by Bailey et al. [181], Gill et al. [182], Jin and Hildebrandt [183], Biju et al. [184] and Kuang et al. [185]. Insights into the analytical potential of QDs focussing on their utilization in automated flow−based and flow-related approaches are summarized in a review by Frigerio et al. [186].

CdSe−ZnS core−shell QDs were used as luminescent labels for microtitre plate assay (fluorescence-linked immunosorbent assay FLISA) towards the detection of various analytes in different matrices. Examples are sulphamethazine [187] and enrofloxacin [188] in chicken muscle tissue; chlorpyrifos in drinking water [189]; *L. monocytogenes* cell surface proteins [190]; multiple detection of dexamethason, gentamicin, clonazepam, medroxyprogesterone acetate and ceftiofur [191]; simultaneous visual detection of multiple antibiotic residues in milk [192]; clenbuterol in pig urine [193] and progesterone in bovine milk [194]. Comparison of the analytical parameters of ELISA and FLISA showed a fourfold decrease in IC_{50} (0.4 and 0.1 ng/mL for zearealenone detection with ELISA and FLISA, respectively) [195]. Even a specific name for QD-based immunoassay was suggested: QLISA [196]. QDs' luminescence was successfully combined with enzyme chemiluminescent labels for simultaneous detection of three cancer markers in human serum [197]. Introduction of QDs into microspheres enhances their brightness and photostability and widens the bioconjugation possibilities, as was reviewed by Ma et al. [198]. Similar results were obtained through incorporation of multiple hydrophobic or hydrophilic QDs into liposomes [61,199].

QDs applied as a label for LFIA was patented for the detection and quantification of one or more analytes [200,201]. According to Web of Science, no manuscripts were published on LFIA with QDs as a label prior to 2009, only some preliminary results were reported. Nowadays, QDs as labels for LFIA were described for trichloropyridinol [202], for the protein biomarker nitrated ceruloplasmin [203], for syphilis antigen [204], α-fetoprotein [205], chloramphenicol in milk [206], foodborne pathogens [207], test for DNA hybridization assays and genotyping of single-nucleotide polymorphisms [208].

A QD-based immunochromatographic test was developed for the simultaneous detection of several compounds in a complex sample matrix. The system was designed in a 'traffic light' format comprising three lines of different colours on a test strip, thereby providing an easy tool with which to identify an analyte of interest based on the visible colour of the line formed (qualitative analysis), and to determine the amount of the analytes (antibiotics ofloxacin, chloramphenicol and streptomycin in milk) present based on the fluorescence intensity of the lines (quantitative analysis). The visually detected cut-off values were 80—200 times lower than those achievable with ELISA using the same antibodies [209].

Comparison of the LFIA's sensitivity with the same set of immunoreagents showed that the LOD of naked-eye CdTe QDs test strips (under a portable UV lamp) was 10 times lower than that of the colloidal gold test strips [204,205].

Similarly, the column gel—based immunotests' sensitivity for benzo[a]pyrene detection in water showed a lower detection limit of 5 ng/L when using HRP or CdSe QDs as a label, but 25 ng/L when using AuNPs. Assays with particle labels required only four consecutive working steps, while those based on HRP required five additions of reagent [13]. Column QD—based immunotests with polyethylene frits and sepharose gel were used for zearalenone detection in wheat [195].

6.2 Up-converting Emitters

UCP utilize a combination of absorber and emitter ions in submicron size crystals (about 200—400 nm in diameter). The absorber ion (energy donor) is excited by the energy of an infrared light source (usually a 980-nm diode laser) and this energy is transferred nonradiatively to the emitter ion (energy acceptor) which radiates a photon in the visible to near infrared range (400—800 nm) depending on the ion composition of the crystal. This anti-Stokes' photoluminescence is based on the sequential absorption of two low energy photons. In contrast to conventional two-photon excitation, the simultaneous absorption of photons is not necessary for up-converting processes. It could occur in the microsecond time scale because of the long lifetime of the metastable excited states [7].

This enables the excitation of UCP by low-energy infrared photons, produced by a laser with relatively low power. In such conditions the photodegradation of biomolecules and the luminescent background are no crucial factors. This technology is also compatible with optically difficult samples, such as whole blood. Because up-conversion processes are unique in regard to the lack of background luminescence from the biochemical assay, there is no need for time-resolved detection. Since the emission wavelengths differ distinctly from the excitation wavelength in the anti-Stokes area, the detection process can be simplified and thus be robust with respect to environmental sampling conditions [210,211]. Inorganic materials capable of up-conversion consist of a host lattice embedded with certain rare earth dopant ions. So far, the most efficient up-converting materials reported have been based on hexagonal sodium yttrium fluoride ($NaYF_4$) host lattices. The dopant ions in the

UCPs are typically trivalent lanthanides which possess multiple, long-lifetime excited states [212]. Different combinations of rare-earth emitting (erbium, holmium, thulium) and absorbing (ytterbium, erbium, samarium) ions create 20 or more unique UCP compositions [213]. The optical properties of the UCP particles are completely unaffected by their environment, since the energy transformation process occurs within the host crystal [210]. In addition, UCP have a narrow emission band, long fluorescent lifetime, high chemical stability and low potential cytotoxicity [214].

A simple way to obtain submicron-sized UCP particles is by bead-milling commercially available bulk phosphor material (Orasure Technologies, Inc.; Phosphor Technology Ltd) and fractionating the smaller particles by sedimentation or flow filtration, based on multiple filter membranes with decreasing pore sizes. In this way, irregular shapes still remain a problem, especially in heterogeneous bioanalytical assays. Large size distribution and nonspecific binding of UCP particles can cause a high assay variation in the analyte's lower concentration range and thus restrict the analytical detection limit. To improve homogeneity in both size and shape of UCP particles the homogeneous precipitation method, followed by a series of fluidized bed processes, has been used. Over the past years, several new methods of synthesis were reported [212].

One disadvantage of UCP technology is the relatively low quantum efficiency of the up-conversion process. Also, the submicron size of UCP particles is quite large relative to the typical proteins' size and can affect the specificity and the kinetics of assays because of steric restrictions on the beads' surface. Phosphors with diameters of 200 nm or less address particle size issues to some degree [213]. Another disadvantage in using UCP particles as labels is the nonspecific binding of untreated phosphor particles with biomolecules [211]. Similar to QDs, UCP particles cannot readily conjugate with immunoreagents. Surface functionalization of UCP particles allows not only broadening of biochemical compatibility and minimizing of nonspecific binding, but also increases the stability of UCP particles by enhancement of the aqueous dispersability and reducing the tendency for aggregation [212].

Commonly, coating of the phosphor surface with tetraethylorthosilicate to achieve a thin (5–50 nm) surface layer of silica is performed. Silanization allows introducing functional groups to the surface of the particles which serve as binding sites for the attachment of biomolecules such as antibodies, antigens, biotin and others with standard cross-linking chemistries. As a result of the common silica coating process, the biomolecules' coupling chemistries are identical for different phosphor crystals [210]. An alternative way for bioconjugation of UCP particles is passive coating with poly(acrylic acids) to insert carboxylic acid groups onto the surface of phosphor particles, followed by activation of carboxylated phosphor and coupling of biomolecules [211].

Application of UCP for bioanalytical assay was reviewed in 2014 [215,216]. Various UCP-LFIA strips have been developed for the POC testing, combining UCP as a reporter with the LFIA principle and small and robust optical systems offering accurate quantitative detection. Tests for high-molecular weight analytes based on a sandwich format were reported for *Schistosoma* circulating anodic antigen [217], *E. coli* [218], *Yersinia pestis* [219], respiratory syncytial virus [220], interferon γ [221], pathogens *Streptococcus pneumonia* [222] and *Brucella* [223], hepatitis B surface antibody [224] and nucleic acids [225,226]. Simultaneous detection of two biomarkers in blood samples was realized with LFIA for the diagnosis of mycobacterial diseases and showed good correlation with ELISA [227]. A competitive assay format was applied for drugs of abuse detection in saliva samples [219] and clenbuterol [228].

A reader called UP*link* from Orasure Technologies, Inc. (Bethlehem, PA) has been developed for UCP-based LFIA. The lowest detectable concentration for the UCP particles was 10–100 emitting particles, depending on their properties [219]. Also, a standard epifluorescence microscope was adapted for visualizing antigens in tissue sections or on cell membranes by eye through the eyepieces [229] using UCP particles.

Several investigations were performed to compare the sensitivity of UCP particles as labels with other reporter systems as well as the sensitivity of UCP-LFIA with other immunochemical methods. Hampl et al. [210] indicated at least a 10-fold increase in sensitivity over conventional LFIA reporter systems such as colloidal gold or coloured latex beads and also ELISA microtitre plate assay. Li et al. [222] reported the best sensitivity of the UCP-LFIA in comparison with the Abbott Axsym AUSAB assay and commercial ELISA test kits.

The availability of UCP with different compositions possessing unique narrow-band emission spectra, all which can be excited by the same light source, allows performing multiple assays. This is not only done by spotting immunoreagents on multiple separate zones, but also within a single zone. Application of two UCP labels for simultaneous detection of two analytes was demonstrated with mouse IgG and ovalbumin (OVA) as targets. The two targets were colour coded using UCP crystals with different compositions: thulium oxysulphide phosphor that emits at 480 nm (blue) and erbium oxysulphide phosphor, which emits at 550 nm (green) [210].

The ability of the UP*link* instrument to detect up to 12 lines on a single assay strip was shown and applied in the detection of a range of narcotics (including amphetamines, methamphetamine, phenylcyclohexylpiperidine and opiates) with a single multiplexed lateral-flow assay strip [218]. Corstjens et al. [230] developed multiplex UCP-LFIA for the detection of human antibodies against human immunodeficiency virus with additional capture zones to detect antibodies against *Myobacterium tuberculosis* or hepatitis C virus. The authors noted that further studies are necessary to evaluate the maximum number of test lines that can be applied to an LFIA strip with special attention to the effect of high signals in the test lines closest to the sample pad.

To improve sensitivity of UCP-based immunoassay UCPs as the signal probe were designed with a core–shell structure which provided a 40-fold enhancement of the luminescence intensity, as reported in [231]. UCPs were covalently conjugated with the anticarcinoembryonic antigen antibody, magnetic beads were combined with another anticarcinoembryonic antigen antibody. With the assistance of a magnet, the as-formed immune sandwich in the presence of CEA can be readily separated from the assay matrix. The immunosensor showed a linear dynamic range 0.1–20 ng/mL in a human serum sample [231].

Luminescence quenching can be mentioned as a perspective for applying UCP as well as other luminescent materials. If a specific antibody is attached to the UCP surface and analytes compete for binding with the analyte-luminescence quenching dye conjugate, this will result in an increase of the luminescent signal of the test zone [232].

6.3 Nanoparticles with Long-Living Emission

In contrast to standard luminescence methods that only use optical filters to separate the emission from the background light through wavelength differences, the time-resolved luminescence techniques separate the emission of interest from the background through lifetime differences. Despite a relatively low intensity in comparison with conventional luminescence, time-resolved luminescence detection techniques have potentially higher sensitivity because of lower background noise. Time resolution–based techniques involve exciting a long-lived emitter with a short pulse of light followed by a waiting period to allow the decay of unwanted emission to a low level, before collecting the remaining long-lived signal [233]. The drawback of this technology is that only a limited number of probes have been discovered which possess sufficient emission lifetime to differentiate significantly from the typical background fluorescence. Time-resolved reporters usually have a decay time of several hundreds of microseconds, optimally >500 μs. This is far more than that of conventional fluorescent probes or autofluorescent samples, typically having decay times of <50 ns. So the autofluorescence of blood and serum can also be completely rejected by the time-resolved luminescence measurements. Long emission lifetime is also critical for constructing simple and low-cost instruments for time-resolved measurements [234]. A portable time-resolved luminescence reader that does not use expensive optical filters, which is a must for the conventional fluorescence detection technique, was described by Song and Knotts [233].

The most common labels for time-resolved fluorescence are based on chelates of the lanthanides (mostly europium) because of their high luminescence quantum yield, large Stokes shift, and the most important reason: a long lifetime [7].

Fluorescence quenching due to water molecules usually leads to weak luminescence of lanthanide chelates in aqueous solution. Incorporating these chelates into submicron particles is an advantageous way to obtain bright fluorescence (by multiple loading and decrease of fluorescence quenching by water molecules), chemical stability and simple bioconjugation possibilities [235,236]. For example, up to 7×10^5 europium chelate molecules could be covalently loaded on each silica nanoparticle [237]. This kind of labels is also commercially available (Molecular Probes Inc., Seradyn Inc.).

The lowest detectable amount reported, was 3.3×10^7 particles/mm^2 [238]. Different lifetime and spectral characteristics of lanthanide chelates allow their application in the simultaneous detection of two different analytes in one physical location [237]. The potential of lanthanide time-resolved luminescence to design sensitive and specific immunoassays, techniques for labelling biomolecules with lanthanide chelate tags, microtitre plate–based assays and the application in luminescence microscopy were discussed by Hagan and Zuchner in 2011 [239]. Application of lanthanide-doped nanoprobes for biodetection was reviewed by Tu et al. [240] and Chen et al. [241].

Application of these labels was published for different immunoassay formats, both heterogeneous and homogeneous. Comparison of time-resolved fluoroimmunoassay in microtitre plate format showed no statistically significant differences in sensitivity and specificity with commercial ELISAs for quantitative determination of herpes simplex virus IgG [242]. Avoiding of enzyme using allows [243] to develop an all-in-one dry-reagent time-resolved fluorometric immunoassay with Eu(III) chelate labels which requires minimal liquid handling for the detection of anti-HIV-1 and -2 antibodies. This assay is envisioned to fill the gap between the rapid point-of-care assays and traditional ELISA in terms of complexity and turnaround time, without compromising the performance. Dual-label time-resolved fluoroimmunoassay in microtitre plate format was described for simultaneous detection of α-fetoprotein and hepatitis B virus surface antigen [244], clothianidin and diniconazole [245], sulphamethazine and sulphaquinoxaline [246] in separate wells. LFIA detection is described for detection of eosinophils and neutrophils in whole blood [238], C-reactive protein in serum [233] and ochratoxin A in agro-products [247].

An additional advantage of lanthanide chelates is a high Stokes shift (>150 nm). This enable elimination of very high background fluorescence from membranes measured by a conventional fluorescence system due to the scattering caused by the membrane itself even without application of time-resolved techniques. The stationary mode of excitation and emission registration allows quantification with digital camera or even visual evaluation using a UV source. Chen et al. [248] showed that the detection sensitivity with the therbium chelate-loaded nanoparticles is more than 100-fold as high as that with fluorescein isothiocyanate molecules. Xia et al. [249] compared the sensitivity of LFIA for hepatitis B surface antigen with different labels and reported a 10-fold higher LFIA sensitivity with europium chelate-loaded silica nanoparticles than the colloidal gold–based LFIA (0.03 μg/L and 3.51 μg/L, respectively).

Time-resolved fluoroimmunoassay protocol using magnetic particles for the simultaneous determination of α-fetoprotein and the free β-subunit of human chorionic gonadotropin in human serum [250] and C-peptide and insulin compounds [251] is described with Eu^{3+} and Sm^{3+} chelate labels. The new approach uses magnetic particles as an immobilization matrix and means of separation, while the luminescent europium and samarium chelates are used as probes. The proposed method shows a wide dynamic range, low detection limit, short analytical time and has potential for multiplex quantification of biomolecules in a single experiment.

An ultrasensitive time-resolved fluorescence immunoassay for bisphenol A where AuNPs were modified with specific antibody and thiolated dsDNA-biotin acting as a signal amplifier. In a competitive reaction, the analyte competes with immobilized analyte-protein conjugate on the surface of microtitre plates to bind to the specific antibodies on the surface of the AuNPs. In the next step, a Eu(III)-labelled streptavidin is added to link to the SH-dsDNA-biotin as a tracer. Fluorescence is amplified via both the AuNPs and the biotin-streptavidin systems, and its intensity is measured in a time-resolved fluorescence immunoassay [252].

Rapid and sensitive homogenous time-resolved fluorescence immunoassays based on the energy-transfer process of europium cryptate donor to AlexaFluor647 acceptor was developed for the detection of both singled-epitope (proteins) and multiepitope (particles) as rapid, simple and sensitive method that can be used to obtain fast results to allow an effective medical care. The energy-transfer process is limited to $d < 10$ nm, making the homogeneous assay an ideal approach assay for examination of homogenous and complex samples, since only mutual binding of the donor and acceptor antibodies to the analyte would result in positive signal. Homogenous time-resolved fluoroimmunoassay was developed for the detection of the bacterial protective antigen toxin, a serological marker that correlates with bacteraemia in infected hosts, using two monoclonal specific antibodies that specifically recognize two different epitopes on the protective antigen molecule. Additionally, homogenous time-resolved fluoroimmunoassay was developed for the detection of bacterial spores using polyclonal antispore antibodies [253].

Another type of time-resolved reporters is phosphorescent chelates. Platinum, palladium and ruthenium chelates possess lifetimes of several 100 μs. Ruthenium complexes excite in the blue range and emit in the red range (600–700 nm). Application of phosphorescence for time-resolved measurements demands deoxygenating of media because of very effective oxygen quenching of long-lived triplet states. One way to prevent oxygen quenching is to encapsulate the phosphorescent molecules in an oxygen-low or oxygen-free matrix. For example, phosphorescent molecules have been encapsulated inside polyacrylonitrile, polystyrene and Sephadex particles. In 2008, halogen-containing polymers and copolymers were found to be excellent encapsulating matrices for phosphorescent chelates to provide phosphorescent nanoparticles with surface functional groups. The sensitivity of LFIA with two kinds of time-resolved labels (commercial europium chelate loaded nanoparticles and synthesized phosphorescent nanoparticles) was compared by Song and Knotts [233]. They reported that europium-based labels were capable of detecting 0.2 ng/mL C-reactive protein in serum. Time-resolved phosphorescence from phosphorescent nanoparticles incorporated in a lateral-flow device was also measured by an in-house developed time-resolved luminescence reader [233].

7. MAGNETIC NANOPARTICLES

MNPs as carriers and labels have been widely used in biochemical separation and detection, blood detoxification and drug delivery [254,255]. Application of MNPs as labels for rapid tests eliminates some disadvantages of optical and noninstrumental visual detection. As support materials are usually not entirely transparent for light in visible and near-UV spectral areas, the signal can only be collected from the top layer of the support. For example, for a nitrocellulose membrane (the common solid support for LFIA), only the signal coming from the top 10 μm of the membrane (usually 100 μm thick) can be detected [256]. In contrast, all magnetic signals originating from the MNPs within the entire volume of the membrane can be detected by the use of a highly sensitive magnetic particle detector. Since none of the generated signal is lost, this improves the sensitivity of the tests. Furthermore, MNP application is associated with a low background noise because extremely low magnetic signals can be detected for most of the biological samples [256]. Unlike other kinds of nanoparticles, the signals generated by MNPs do not degrade over time. Another common feature is that the size of the MNPs largely determines its flow rate and affects the detection time. The magnetite content of MNPs is positively correlated to the signal intensity so an increasing magnetite content displays a stronger signal [256]. With these advantages, MNPs can be a perspective label for clinical research, environmental and food safety testing. For results' interpretation the portable Magnetic Assay Reader (MAR) (MagnaBioScience, Quantum Design, San Diego) was developed. The tendency to associate in solutions because of their magnetic properties can be mentioned as a disadvantage [7].

MNPs conjugation with proteins is commonly performed by the synthesis of iron oxide (either Fe_3O_4 or γ-Fe_2O_3) nanoparticles, silica-coating, surface modification with amino or carboxylic groups, and conjugation with biomolecules [257]. Some examples of magnetic-based LFIA are developed for clinical applications, such as tests for hCG hormone [257], cardiac troponin I [258], *E. coli* [259] human papillomavirus [260] and HIV antibodies [261].

ACKNOWLEDGEMENT

The work was supported by Russian Science Foundation (project 14-13-00229).

REFERENCES

[1] B. O'Farrell, Lateral flow immunoassay systems: evolution from the current state of the art to the next generation of highly sensitive, quantitative rapid assays, in: D. Wild (Ed.), The Immunoassay Handbook Theory and Applications of Ligand Binding, ELISA and Related Techniques, fourth ed., Elsevier, 2013.

[2] G.A. Posthuma-Trumpie, J. Korf, A. van Amerongen, Lateral flow (immuno)assay: its strengths, weaknesses, opportunities and threats. A literature survey, Anal. Bioanal. Chem. 393 (2009) 569–582.

[3] B.B. Dzantiev, N.A. Byzova, A.E. Urusov, A.V. Zherdev, Immunochromatographic methods in food analysis, Trends Anal. Chem. 55 (2014) 81–93.

[4] A. Warsinke, Point-of-care testing of proteins, Anal. Bioanal. Chem. 393 (2009) 1393–1405.

[5] M. Seydack, Nanoparticle labels in immunosensing using optical detection methods, Biosens. Bioelectron. 12 (2005) 2454–2469.

[6] C. Cháfer-Pericás, A. Maquieira, R. Puchades, Functionalized inorganic nanoparticles used as labels in solid-phase immunoassays, Trends Anal. Chem. 31 (2012) 144–156.

[7] I.Y. Goryacheva, P. Lenain, S. De Saeger, Nanosized labels for rapid immunotests: new trends and developments, Trends Anal. Chem. 46 (2013) 30–46.

[8] I.Y. Goryacheva, E.S. Speranskaya, V.V. Goftman, D. Tang, S. De Saeger, Synthesis and bioanalytical applications of nanostructures multiloaded with quantum dots, Trends Anal. Chem. 66 (2015) 53–62.

[9] D. Pyo, J. Yoo, New trends in fluorescence immunochromatography, J. Immunoassay. Immunochem. 33 (2012) 203–222.

[10] I. Weeks, L.J. Kricka, D. Wild, Signal generation and detection systems (excluding homogeneous assays), in: D. Wild (Ed.), The Immunoassay Handbook Theory and Applications of Ligand Binding, ELISA and Related Techniques, fourth ed., Elsevier, 2013.

[11] G.T. Hermanson, Bioconjugate Techniques, third ed., Elsevier, 2013, p. 1146.

[12] B. Zhang, D. Tang, I.Y. Goryacheva, R. Niessner, D. Knopp, Anodic stripping voltammetric immunoassay for ultrasensitive detection of low-abundance-proteins using quantum dots-aggregated hollow microspheres, Chem. Eur. J. 19 (2013) 2496–2503.

[13] N.V. Beloglazova, I.Y. Goryacheva, R. Niessner, D. Knopp, A comparison of horseradish peroxidase, gold nanoparticles and quantum dots as labels in non-instrumental gel-based immunoassay, Microchim. Acta 175 (2011) 361–367.

[14] S.M. Han, J.H. Cho, I.H. Cho, E.H. Paek, H.B. Oh, B.S. Kim, C. Ryu, K. Lee, Y.K. Kim, S.H. Paek, Plastic enzyme-linked immunosorbent assays (ELISA)-on-a-chip biosensor for botulinum neurotoxin A, Anal. Chim. Acta 587 (2007) 1–8.

[15] I.H. Cho, E.H. Paek, Y.K. Kim, J.H. Kim, S.H. Paek, Chemiluminometric enzyme-linked immunosorbent assays (ELISA)-on-a-chip biosensor based on cross-flow chromatography, Anal. Chim. Acta 632 (2009) 247–255.

[16] C. Parolo, A. de la Escosura-Muniz, A. Merkoci, Enhanced lateral flow immunoassay using gold nanoparticles loaded with enzymes, Biosens. Bioelectron. 40 (2013) 412–416.

[17] Y. He, S. Zhang, X. Zhang, M. Baloda, A.S. Gurung, H. Xu, X. Zhang, G. Liu, Ultrasensitive nucleic acid biosensor based on enzyme–gold nanoparticle dual label and lateral flow strip biosensor, Biosens. Bioelectron. 26 (2011) 2018–2024.

[18] S. Binder, J.A. Isler, Detection of antibodies relevant to infectious disease, in: D. Wild (Ed.), The Immunoassay Handbook Theory and Applications of Ligand Binding, ELISA and Related Techniques Fourth Edition, Elsevier, 2013.

[19] C. Zhang, Y. Zhang, S. Wang, Development of multianalyte flow-through and lateral-flow assays using gold particles and horseradish peroxidase as tracers for the rapid determination of carbaryl and endosulfan in agricultural products, J. Agric. Food Chem. 54 (2006) 2502–2507.

[20] K.K. Fung, C.P.Y. Chan, R. Renneberg, Development of a creatinine enzyme-based bar-code-style lateral-flow assay, Anal. Bioanal. Chem. 393 (2009) 1281–1287.

[21] K.K. Fung, C.P.Y. Chan, R. Renneberg, Development of enzyme-based bar code-style lateral-flow assay for hydrogen peroxide determination, Anal. Chim. Acta 634 (2009) 89–95.

[22] J.H. Cho, S.M. Han, E.H. Paek, I.H. Cho, S.H. Paek, Plastic ELISA-on-a-chip based on sequential cross-flow chromatography, Anal. Chem. 78 (2006) 793–800.

[23] I.H. Cho, J. Irudayaraj, Lateral-flow enzyme immunoconcentration for rapid detection of *Listeria monocytogenes*, Anal. Bioanal. Chem. 405 (2013) 3313–3319.

[24] H.A. Joung, Y.K. Oh, M.G. Kim, An automatic enzyme immunoassay based on a chemiluminescent lateral flow immunosensor, Biosens. Bioelectron. 53 (2014) 330–335.

[25] M. Mirasoli, A. Buragina, L.S. Dolci, M. Guardigli, P. Simoni, A. Montoyac, E. Maiolini, S. Girotti, A. Roda, Development of a chemiluminescence-based quantitative lateral flow immunoassay for on-field detection of 2,4,6-trinitrotoluene, Anal. Chim. Acta 721 (2012) 167–172.

[26] M. Zangheri, L. Cevenini, L. Anfossi, C. Baggiani, P. Simoni, F. Di Nardo, A. Roda, A simple and compact smartphone accessory for quantitative chemiluminescence-based lateral flow immunoassay for salivary cortisol detection, Biosens. Bioelectron. 64 (2015) 63–68.

[27] S.J. Wang, W.F. Chang, M.Y. Wang, K.P. Hsiung, Y.C. Liu, Development of a disperse dye immunochromatographic test for the detection of antibodies against infectious bursal disease virus, Vet. Immunol. Immunopathol. 125 (2008) 284–290.

[28] Y. Zhu, W. He, Y. Liang, M. Xu, C. Yu, W. Hua, G. Chao, Development of a rapid, simple dipstick dye immunoassay for schistosomiasis diagnosis, J. Immunol. Methods 266 (2002) 1–5.

[29] S. Jin, Z.Y. Chang, X. Ming, C.L. Min, H. Wei, L.Y. Sheng, H.G. Xiao, Fast dipstick dye immunoassay for detection of immunoglobulin G (IgG) and IgM antibodies of human toxoplasmosis, Clin. Diagn. Lab. Immunol. 12 (2005) 198–201.

[30] X. Xiang, W. Tianping, T. Zhigang, Development of a rapid, sensitive, dye immunoassay for schistosomiasis diagnosis: a colloidal dye immunofiltration assay, J. Immunol. Methods 280 (2003) 49–57.

[31] I.A. Lubavina, I.S. Salomatina, A.A. Zinchenko, A.V. Zherdev, B.B. Dzantiev, A noninstrumental immunoassay based on colloidal dyes, Russ. J. Bioorg. Chem. 26 (2000) 207–212.

[32] K. Snowden, M. Hommel, Antigen detection immunoassay using dipsticks and colloidal dyes, J. Immunol. Methods 140 (1991) 57–65.

[33] W.F. Chang, S.J. Wang, S.F. Lai, C.J. Shieh, K.P. Hsiung, Y.C. Liu, Evaluation on the use of reactive dye-modified polylysine as the biomarker in immunochromatographic test application, Anal. Biochem. 411 (2011) 236–240.

[34] W.C. Mak, K.K. Sin, C.P.Y. Chan, L.W. Wong, R. Renneberg, Biofunctionalized indigo-nanoparticles as biolabels for the generation of precipitated visible signal in immunodipsticks, Biosens. Bioelectron. 26 (2011) 3148–3153.

[35] S. Ahn-Yoon, T.R. DeCory, R.A. Durst, Ganglioside–liposome immunoassay for the detection of botulinum toxin, Anal. Bioanal. Chem. 378 (2004) 68–75.

[36] K.A. Edwards, A.J. Baeumner, Optimization of DNA-tagged dye-encapsulating liposomes for lateral-flow assays based on sandwich hybridization, Anal. Bioanal. Chem. 386 (2006) 1335–1343.

[37] J.A.A. Ho, R.D. Wauchope, A strip liposome immunoassay for aflatoxin B1, Anal. Chem. 74 (2002) 1493–1496.

[38] H.W. Wen, W. Borejsza-Wysocki, T.R. DeCory, R.A. Durst, Development of a competitive liposome-based lateral flow assay for the rapid detection of the allergenic peanut protein Ara h1, Anal. Bioanal. Chem. 382 (2005) 1217–1226.

[39] Y.M. Kim, S.W. Oh, S.Y. Jeong, D.J. Pyo, E.Y. Choi, Development of an ultrarapid one-step fluorescence immunochromatographic assay system for the quantification of microcystins, Environ. Sci. Technol. 37 (2003) 1899–90437.

[40] S.W. Oh, Y.M. Kim, H.J. Kim, S.J. Kim, J.S. Cho, E.Y. Choi, Point-of-care fluorescence immunoassay for prostate specific antigen, Clin. Chim. Acta 406 (2009) 18–22.

[41] S.W. Oh, J.D. Moon, S.Y. Park, H.J. Jang, J.H. Kim, K.B. Nahm, E.Y. Choi, Evaluation of fluorescence hs-CRP immunoassay for point-of-care testing, Clin. Chim. Acta 356 (2005) 172–177.

[42] S. Choi, E.Y. Choi, D.J. Kim, J.H. Kim, T.S. Kim, S.W. Oh, A rapid, simple measurement of human albumin in whole blood using a fluorescence immunoassay (I), Clinica Chim. Acta 339 (2004a) 147–156.

[43] S. Choi, E.Y. Choi, H.S. Kim, S.W. Oh, On-site quantification of human urinary albumin by a fluorescence immunoassay, Clin. Chem. 50 (2004b) 1052–1055.

[44] D. Knopp, D. Tang, R. Niessner, Review: bioanalytical applications of biomolecule-functionalized nanometer-sized doped silica particles, Anal. Chim. Acta 647 (2009) 14–30.

[45] R.I. Nooney, E. McCormack, C. McDonagh, Optimization of size, morphology and colloidal stability of fluorescein dye-doped silica NPs for application in immunoassays, Anal. Bioanal. Chem. 404 (2012) 2807–2818.

[46] S.W. Bae, W. Tan, J.I. Hong, Fluorescent dye-doped silica nanoparticles: new tools for bioapplications, Chem. Commun. 48 (2012) 2270–2282.

[47] W. Yang, C.G. Zhang, H.Y. Qu, H.H. Yang, J.G. Xu, Novel fluorescent silica nanoparticle probe for ultrasensitive immunoassays, Anal. Chim. Acta 503 (2004) 163–169.

[48] S. Renshaw, Immunohistochemistry and immunocytochemistry, in: D. Wild (Ed.), The Immunoassay Handbook Theory and Applications of Ligand Binding, ELISA and Related Techniques, fourth ed., Elsevier, 2013.

[49] W. Yang, D. Trau, R. Renneberg, N.T. Yu, F. Caruso, Layer-by-layer construction of novel biofunctional fluorescent microparticles for immunoassay applications, J. Colloid Interf. Sci. 234 (2001) 356–362.

[50] D. Trau, W. Yang, M. Seydack, F. Caruso, N.T. Yu, R. Renneberg, Nanoencapsulated microcrystalline particles for superamplified biochemical assays, Anal. Chem. 74 (2002) 5480–5486.

[51] C.P.Y. Chan, Y. Bruemmel, M. Seydack, K.-K. Sin, L.W. Wong, E. Merisko-Liversidge, D. Trau, R. Renneberg, Nanocrystal biolabels with releasable fluorophores for immunoassays, Anal. Chem. 76 (2004) 3638–3645.

[52] Y. Bruemmel, C.P.Y. Chan, R. Renneberg, A. Thuenemann, M. Seydack, On the influence of different surfaces in nano- and submicrometer particle based fluorescence immunoasays, Langmuir 20 (2004) 9371–9379.

[53] S. Fery-Forgues, Fluorescent organic nanocrystals and non-doped nanoparticles for biological applications, Nanoscale 5 (2013) 8428–8442.

[54] A.D. Bangham, M.M. Standish, J.C. Watkins, Diffusion of univalent ions across the lamellae of swollen phospholipids, J. Mol. Biol. 13 (1965) 238–252.
[55] H.A.H. Rongen, A. Bult, W.P. Van Bennekom, Liposomes and immunoassays, J. Immunol. Methods 204 (1997) 105–133.
[56] F. Szoka, D. Papahadjopoulos, Procedure for preparation of liposomes with large internal aqueous space and high capture by reverse-phase evaporation, Proc. Natl. Acad. Sci. U.S.A. 75 (1978) 4194–4198.
[57] C. Pidgeon, A.H. Hunt, K. Dittrich, Formation of multilayered vesicles from water/organic-solvent (W/O) emulsions: theory and practice, Pharmaceut. Res. 3 (1986) 23–34.
[58] W. Jiskoot, T. Teerlink, E.C. Beuvery, D.J.A. Crommelin, Preparation of liposomes via detergent removal from mixed micelles by dilution. The effect of bilayer composition and process parameters on liposome characteristics, Pharm. Weekbl. Sci. Ed. 8 (1986) 259–265.
[59] R.R.C. New (Ed.), Liposome. A Practical Approach, Oxford University Press, Oxford, 1990.
[60] M.A. Jones, P.K. Kilpatrick, R.G. Carboriell, Preparation and characterization of bifunctional unilamellar vesicles for enhanced immunosorbent assays, Biotechnol. Prog. 9 (1993) 242–258.
[61] N.V. Beloglazova, P.S. Shmelin, E.S. Speranskaya, B. Lucas, C. Helmbrecht, D. Knopp, R. Niessner, S. De Saeger, I.Y. Goryacheva, Quantum dot loaded liposomes as new luminescent labels for immunoassay, Anal. Chem. 85 (2013) 7197–7204.
[62] J.R. Nordlund, C.F. Schmidt, S.N. Dicken, T.E. Thompson, Biochemistry 20 (1981) 3237–3241.
[63] J.M. Kremer, M.W. Esker, C. Pathmamanoharan, P.H. Wiersema, Biochemistry 16 (1977) 3932–3935.
[64] D.D. Lasic, D. Papahadjopoulos, Liposomes revisited, Science 267 (1995) 1275–1276.
[65] G. Gregoriadis, A.T. Florence, Liposomes in drug delivery. Clinical, diagnostic and ophthalmic potential, Drugs 45 (1993) 15–28.
[66] J.G. Smith, R.L. Walzem, J.B. German, Liposomes as agents of DNA transfer, Biochim. Biophys. Acta 1154 (1993) 327–340.
[67] K.A. Edwards, A.J. Baeumner, Liposomes in analyses, Talanta 68 (2006) 1421–1431.
[68] A. Gomez-Hens, J.M. Fernandez-Romero, The role of liposomes in analytical processes, Trends Anal. Chem. 24 (2005) 9–19.
[69] Y. Ishimori, T. Yasuda, T. Tsumita, M. Notsuki, M. Koyama, T. Tadakuma, Liposome Immune Lysis Assay (LILA); a simple method to measure anti-protein antibody using protein antigen-bearing liposomes, J. Immunol. Methods 75 (1984) 351–360.
[70] R.J. Ho, L. Huang, Immunoliposome assays: perspectives, progress and potential, in: G. Gregoriadis (Ed.), Liposomes as Drug Carriers, John Wiley and Sons, New York, 1988, pp. 527–547.
[71] [a] B.S. Yu, Y.K. Choi, H. Chung, Development of immunoassay methods by use of liposomes, Biotechnol. Appl. Biochem. 9 (1987) 209–216;
[b] F.J. Martin, V.T. Kueg, Use of liposomes as agglutination enhancement agents in diagnostic test, Methods Enzymol. 149 (1987) 200–213.
[72] K. Tomioka, W. Okada, H. Fukuda, S. Katoh, Peptide-coupled liposomes for homogeneous immunoassay of polyclonal antibodies against proteins, Chem. Eng. J. 62 (1996) 169–174.
[73] Y. Zheng, H. Chen, X.-P. Liu, J.H. Jiang, Y. Luo, G.L. Shen, R.Q. Yu, An ultrasensitive chemiluminescence immunosensor for PSA based on the enzyme encapsulated liposome, Talanta 77 (2008) 809–814.

[74] J.A.A. Ho, H.W. Hsu, M.R. Huang, Liposome-based microcapillary immunosensor for detection of *Escherichia coli* O157:H7, Anal. Biochem. 330 (2004) 342–349.

[75] H. Chen, Y. Zheng, J. Jiang, H. Wu, G. Shen, R. Yu, An ultrasensitive chemiluminescence biosensor for cholera toxin based on ganglioside-functionalized supported lipid membrane and liposome, Biosens. Bioelectron. 24 (2008) 684–689.

[76] K.A. Edwards, Y. Wang, A.J. Baeumner, Aptamer sandwich assays: human α-thrombin detection using liposome enhancement, Anal. Bioanal. Chem. 398 (2010) 2645–2654.

[77] R. Wiese, Analysis of several fluorescent detector molecules for protein assay use, Luminescence 18 (2003) 25–30.

[78] A.J. Baeumner, N.A. Schlesinger, N.S. Slutzki, J. Romano, E.M. Lee, R.A. Montagna, Biosensor for dengue virus detection: sensitive, rapid, and serotype specific, Anal Chem. 74 (2002) 1442–1448.

[79] N.V. Zaytseva, R.A. Montagna, E.M. Lee, A.J. Baeumner, Multi-analyte single-membrane biosensor for the serotype-specific detection of dengue virus, Anal. Bioanal. Chem. 380 (2004) 46–53.

[80] J.A.A. Ho, L.C. Wu, L.H. Chang, K.C. Hwang, J.R.R. Hwu, Liposome-based immunoaffinity chromatographic assay for the quantitation of immunoglobulin E in human serum, J. Chromatogr. B 878 (2010) 172–176.

[81] J.A.A. Ho, S.C. Zeng, W.H. Tseng, Y.J. Lin, C.H. Chen, Liposome-based immunostrip for the rapid detection of Salmonella, Anal. Bioanal. Chem. 391 (2008) 479–485.

[82] J.A.A. Ho, M.R. Huang, Application of a liposomal bioluminescent label in the development of a flow injection immunoanalytical system, Anal. Chem. 77 (2005) 3431–3436.

[83] E. Ralston, L.M. Hjelmeland, R.D. Klausner, J.N. Weinstein, R. Blumenthal, Carboxyfluorescein as a probe for liposome-cell interactions. Effect of impurities, and purification of the dye, Biochim. Biophys. Acta 649 (1981) 133–137.

[84] P.I. Lelkes, Methodological aspects dealing with stability measurements of liposomes in vitro using the carboxyfluorescein-assay, in: G. Gregoriadis (Ed.), Liposome Technology, vol. III, CRC Press, Boca Raton, FL, 1984, p. 225 (Chapter 15).

[85] T.M. Allen, Calcein as a tool in liposome methodology, in: G. Gregoriadis (Ed.), Liposome Technology, vol. III, CRC Press, Boca Raton, FL, 1984, p. 177 (Chapter 12).

[86] J.N. Israelachvili, D.J. Mitchell, A model for the packing of lipids in bilayer membranes, Biochim. Biophys. Acta 389 (1975) 13–19.

[87] T.D. Heath, D. Robertson, M.S.C. Birbeck, A.J.S. Davies, Covalent attachment of horseradish peroxidase to the outer surface of liposomes, Biochim. Biophys. Acta 599 (1980) 42–62.

[88] D.B. Wagner, R.A. Baffi, Vesicle Including a Metal Marker for Use in an Assay, US. Patent No. 4,707,453, November 17, 1987.

[89] T. Ruysschaert, M. Germain, J.F. Pereira da Silva Gomes, D. Fournier, G.B. Sukhorukov, W. Meier, M. Winterhalter, Liposome-based nanoparticles, IEEE Trans. Nanobiosci. 3 (2004) 49–55.

[90] M. Ciobanu, B. Heurtault, P. Schultz, C. Ruhlmannb, C.D. Muller, B. Frisch, Layersome: development and optimization of stable liposomes as drug delivery system, Int. J. Pharm. 344 (2007) 154–157.

[91] J. Hotz, W. Meier, Polymer particles by templating of vesicles, Adv. Mater. 10 (1998) 1387–1390.

[92] M. Jung, D.H.W. Hubert, P.H.H. Bomans, P. Frederik, A.M. van Herk, A.L. German, A topology map for novel vesicle–polymer hybrid architectures, Adv. Mater. 12 (2000) 210–213.

[93] V.J. Mohanraj, T.J. Barnes, C.A. Prestidge, Silica nanoparticle coated liposomes: a new type of hybrid nanocapsule for proteins, Int. J. Pharm. 392 (2010) 285–293.
[94] S. Begu, A.A. Pouessel, D.A. Lerner, C. Tourne-Peteilh, J.M. Devoisselle, Liposil, a promising composite material for drug storage and release, J. Control. Release 118 (2007) 1–6.
[95] N.V. Beloglazova, O.A. Goryacheva, E.S. Speranskaya, T. Aubert, I.Y. Goryacheva, S. De Saeger, Silica-coated liposomes loaded with quantum dots as labels for multiplex fluorescent immunoassay, Talanta 134 (2015) 120–125.
[96] M.A. Jones, A. Singh, P.K. Kilpatrick, R.G. Carbonell, Preparation and characterization of ligand-modified labeled liposomes for solid phase immunoassays, J. Liposome Res. 3 (1993) 793–804.
[97] G.P. Vonk, D.B. Wagner, Encapsulation of europium in liposomes for use in an amplified immunoassay detection system, Clin. Chem. 37 (1991) 1519–1520.
[98] I.M. Roitt, J. Brostoff, D.K. Male, Immunology, third ed., Gower Medical Publishing, London, 1993.
[99] S.C. Kinsky, R.A. Nicolotti, Immunological properties of model membranes, Ann. Rev. Biochem. 46 (1977) 49–67.
[100] J. Szebeni, L. Baranyi, S. Savay, J. Milosevits, M. Bodo, R. Bunger, C.R. Alving, The interaction of liposomes with the complement system: in vitro and in vivo assays, Methods Enzymol. Liposomes Part C 373 (2003) 136–154.
[101] S.C. Kinsky, Antibody-complement interaction with lipid model membranes, Biochim. Biophys. Acta 265 (1972) 1–23.
[102] H.R. Six, K. Uemura, S.C. Kinsky, Effect of immunoglobulin class and affinity on the initiation of complement-dependent damage to liposomal model membranes sensitized with dinitrophenylated phospholipids, Biochem. 12 (1973) 4003–4011.
[103] H.R. Six, W.W. Young Jr., K. Uemura, S.C. Kinsky, Effect of antibody-complement on multiple vs. single compartment liposomes. Application of a fluorometric assay for following changes in liposomal permeability, Biochem. 13 (1974) 4050–4058.
[104] M. Umeda, Y. Ishimori, K. Koshikawa, M. Takada, T. Yasuda, Homogeneous determination of C-reactive protein in serum using liposome immune lysis assay (LILA), Jpn. J. Exp. Med. 56 (1986) 35–42.
[105] M. Umeda, Y. Ishimori, K. Koshikawa, M. Takada, T. Yasuda, Liposome immune lysis assay (LILA). Application of sandwich method to determine a serum protein component with antibody-bearing liposomes, J. Immunol. Methods 95 (1986) 15–21.
[106] K. Otoda, S. Kimura, Y. Imanishi, Interaction of mellitin derivatives with lipid bilayer membrane. Role of basic residues at the C-terminal and their replacement with lactose, Biochim. Biophys. Acta 1112 (1992) 1–6.
[107] W.J. Litchfield, J.W. Freytag, M. Adamich, Highly sensitive immunoassays based on use of liposomes without complement, Clin. Chem. 30 (1984) 1441–1445.
[108] F.S. Ligler, R. Bredehorst, A. Talebian, L.C. Shriver, C.F. Hammer, J.P. Sheridan, C.W. Vogel, B.P. Gaber, A homogeneous immunoassay for the mycotoxin T-2 utilizing liposomes, monoclonal antibodies, and complement, Anal. Biochem. 163 (1987) 369–375.
[109] S. Zhang, Z. Liu, N. Zhou, Z. Wang, J. Shen, A liposome immune lysis assay for enrofloxacin in carp and chicken muscle, Anal. Chim. Acta 612 (2008) 83–88.
[110] L. Dykman, N. Khlebtsov, Gold nanoparticles in biomedical applications: recent advances and perspectives, Chem. Soc. Rev. 41 (2012) 2256–2282.
[111] J.E. Beesley, Colloidal gold: a new revolution in marking cytochemistry, Proc. R. Microsc. Soc. 20 (1985) 187–197.

[112] W.P. Faulk, G.M. Taylor, An immunocolloid method for the electron microscope, Immunochemistry 8 (1971) 1081–1083.

[113] N. Khlebtsov, V. Bogatyrev, L. Dykman, B. Khlebtsov, S. Staroverov, A. Shirokov, L. Matora, V. Khanadeev, T. Pylaev, N. Tsyganova, G. Terentyuk, Analytical and theranostic applications of gold nanoparticles and multifunctional nanocomposites, Theranostics 3 (2013) 167–180.

[114] N. Khlebtsov, L. Dykman, Biodistribution and toxicity of engineered gold nanoparticles: a review of in vitro and in vivo studies, Chem. Soc. Rev. 40 (2011) 1647–1671.

[115] Z. Wang, L.N. Ma, Gold nanoparticle probes, Coord. Chem. Rev. 253 (2009) 1607–1618.

[116] P. Chun, Colloidal gold and other labels for lateral flow immunoassays, in: R.C. Wong, H.Y. Tse (Eds.), Lateral Flow Immunoassay, Humana Press, New York, 2009, pp. 75–93.

[117] L.A. Dykman, V.A. Bogatyrev, Gold nanoparticles: preparation, functionalization and applications in biochemistry and immunochemistry, Russ. Chem. Rev. 76 (2007) 199–213.

[118] P. Bao, A.G. Frutos, C. Greef, J. Lahiri, U. Muller, T.C. Peterson, L. Warden, X. Xie, High-sensitivity detection of DNA hybridization on microarrays using resonance light scattering, Anal. Chem. 74 (2002) 1792–1797.

[119] B. Khlebtsov, N. Khlebtsov, Enhanced solid-phase immunoassay using gold nanoshells: effect of nanoparticle optical properties, Nanotechnology 19 (2008) 435703.

[120] K. Glynou, P.C. Ioannou, T.K. Christopoulos, V. Syriopoulou, Oligonucleotide-functionalized gold nanoparticles as probes in a dry-reagent strip biosensor for DNA analysis by hybridization, Anal. Chem. 75 (2003) 4155–4160.

[121] J. Liu, D. Mazumdar, Y. Lu, A simple and sensitive "dipstick" test in serum based on lateral flow separation of aptamer-linked nanostructures, Angew. Chem. Int. Ed. 45 (2006) 7955–7959.

[122] D. Mazumdar, J. Liu, G. Lu, J. Zhou, Y. Lu, Easy-to-use dipstick tests for detection of lead in paints using non-cross-linked gold nanoparticle–DNAzyme conjugates, Chem. Commun. 46 (2010) 1416–1418.

[123] E. Panfilova, A. Shirokov, B. Khlebtsov, L. Matora, N. Khlebtsov, Multiplexed dot immunoassay using Ag nanocubes, Au/Ag alloy nanoparticles, and Au/Ag nanoboxes, Nano Res. 5 (2011) 124–134.

[124] Y. Ji, M. Ren, Y. Li, Z. Huang, M. Shu, H. Yang, Y. Xiong, Y. Xu, Detection of aflatoxin B1 with immunochromatographic test strips: enhanced signal sensitivity using gold nanoflowers, Talanta 142 (2015) 206–212.

[125] J.Y. Liao, H. Li, Lateral flow immunodipstick for visual detection of aflatoxin B1 in food using immuno-nanoparticles composed of a silver core and a gold shell, Microchim. Acta 171 (2010) 289–295.

[126] D. Tang, J.C. Sauceda, S. Ott, E. Basova, I. Goryacheva, S. Biselli, R. Niessner, D. Knopp, Magnetic nanogold microspheres-based lateral-flow immunodipstick for rapid detection of aflatoxin B2 in food, Biosens. Bioelectron. 25 (2009) 514–518.

[127] H. Xu, J. Chen, J. Birrenkott, J.X. Zhao, S. Takalkar, K. Baryeh, G. Liu, Gold nanoparticle decorated silica nanorods for sensitive visual detection of proteins, Anal. Chem. 86 (2014) 7351–7359.

[128] X. Huang, Z.P. Aguilar, H. Xu, W. Lai, Y. Xiong, Membrane-based lateral flow immunochromatographic strip with nanoparticles as reporters for detection: a review, Biosens. Bioelectron. 75 (2016) 166–180.

[129] L. Scopsi, I. Larsson, L. Bastholm, M.H. Nielsen, Silver-enhanced colloidal gold probes as markers for scanning electron microscopy, Histochemistry 86 (1986) 35–41.

[130] I.H. Cho, S.M. Seo, E.H. Paek, S.H. Paek, Immunogold-silver staining-on-a-chip biosensor based on cross-flow chromatography, J. Chromatogr. B 878 (2010) 271–277.

[131] J.K. Horton, S. Swinburne, M.J. O'Sullivan, A novel, rapid, single-step immunochromatographic procedure for the detection of mouse immunoglobulin, J. Immunol. Methods 140 (1991) 131–134.

[132] M. Yang, C. Wang, Label-free immunosensor based on gold nanoparticle silver enhancement, Anal. Biochem. 385 (2009) 128–131.

[133] C.H. Yeh, C.Y. Hung, T.C. Chang, H.P. Lin, Y.C. Lin, An immunoassay using antibody-gold nanoparticle conjugate, silver enhancement and flatbed scanner, Microfluid. Nanofluid. 6 (2009) 85–91.

[134] W.Y. Wu, Z.P. Bian, W. Wang, W. Wang, J.J. Zhu, PDMS gold nanoparticle composite film-based silver enhanced colorimetric detection of cardiac troponin I, Sens. Actuat. B 147 (2010) 298–303.

[135] B.A. Rohrman, V. Leautaud, E. Molyneux, R.R. Richards-Kortum, A lateral flow assay for quantitative detection of amplified HIV-1 RNA, PLoS One 7 (2012) e45611.

[136] J. Li, M. Zou, Y. Chen, Q. Xue, F. Zhang, B. Li, Y. Wang, X. Qi, Y. Yang, Gold immunochromatographic strips for enhanced detection of Avian influenza and Newcastle disease viruses, Anal. Chim. Acta 782 (2013) 54–58.

[137] N. Nagatani, R. Tanaka, T. Yuhi, T. Endo, K. Kerman, Y. Takamura, E. Tamiya, Gold nanoparticle-based novel enhancement method for the development of highly sensitive immunochromatographic test strips, Sci. Technol. Adv. Mat. 7 (2006) 270–275.

[138] R. Tanaka, T. Yuhi, N. Nagatani, T. Endo, K. Kerman, Y. Takamura, E. Tamiya, A novel enhancement assay for immunochromatographic strips using gold nanoparticles, Anal. Bioanal. Chem. 385 (2006) 414–1420.

[139] Z. Mei, W. Qu, Y. Deng, H. Chu, J. Cao, F. Xue, L. Zheng, H.S. El-Nezamic, Y. Wu, W. Chen, One-step signal amplified lateral flow strip biosensor for ultrasensitive and on-site detection of bisphenol A (BPA) in aqueous samples, Biosens. Bioelectron. 49 (2013) 457–461.

[140] D.H. Choi, S.K. Lee, Y.K. Oh, B.W. Bae, S.D. Lee, S. Kim, Y.B. Shin, M.G. Kim, A dual gold nanoparticle conjugate-based lateral flow assay (LFA) method for the analysis of troponin I, Biosens. Bioelectron. 25 (2010) 1999–2002.

[141] C.F. Duan, Y.Q. Yu, H. Cui, Gold nanoparticle-based immunoassay by using non-stripping chemiluminescence detection, Analyst 133 (2008) 1250–1255.

[142] R. Elghanian, J.J. Storhoff, R.C. Mucic, R.L. Letsinger, C.A. Mirkin, Selective colorimetric detection of polynucleotides based on the distance-dependent optical properties of gold nanoparticles, Science 277 (1997) 1078–1081.

[143] L. Anfossi, C. Baggiani, C. Giovannoli, G. Giraudi, Homogeneous immunoassay based on gold nanoparticles and visible absorption detection, Anal. Bioanal. Chem. 394 (2009) 507–512.

[144] L. Li, B. Li, D. Cheng, L. Mao, Visual detection of melamine in raw milk using gold nanoparticles as colorimetric probe, Food Chem. 122 (2010) 895–900.

[145] L. Guo, J. Zhong, J. Wu, F.F. Fu, G. Chen, X. Zheng, S. Lin, Visual detection of melamine in milk products by label-free gold nanoparticles, Talanta 82 (2010) 1654–1658.

[146] X. Liang, H. Wei, Z. Cui, J. Deng, Z. Zhang, X. You, X.E. Zhang, Colorimetric detection of melamine in complex matrices based on cysteamine-modified gold nanoparticles, Analyst 136 (2011) 179–183.

[147] Y. Zhang, B. Li, X. Chen, Simple and sensitive detection of dopamine in the presence of high concentration of ascorbic acid using gold nanoparticles as colorimetric probes, Microchim. Acta 168 (2010) 107–113.

[148] Y. Zhang, B. Li, C. Xu, Visual detection of ascorbic acid via alkyne–azide click reaction using gold nanoparticles as a colorimetric probe, Analyst 135 (2010) 1579–1584.

[149] W. Zhao, A.M. Monsur, S.D. Aguirre, M.A. Brook, Y. Li, Paper-based bioassays using gold nanoparticle colorimetric probes, Anal. Chem. 80 (2008) 8431–8437.

[150] A. De la Escosura-Muñiz, C. Parolo, A. Merkoç, Immunosensing using nanoparticles, Mater. Today 13 (2010) 17–27.

[151] A. Van Amerongen, J.H. Wichers, L.B. Berendsen, A.J. Timmermans, G.D. Keizer, A.W. van Doorn, A. Bantjes, W.M. van Gelder, Colloidal carbon particles of a new label for rapid immunochemical test methods: quantitative computer image analyses of results, J. Biotechnol. 30 (1993) 185–195.

[152] G.A. Posthuma-Trumpie, J.H. Wichers, M. Koets, L.B.J.M. Berendsen, A. Van Amerongen, Amorphous carbon nanoparticles: a versatile label for rapid diagnostic (immuno) assays, Anal. Bioanal. Chem. 402 (2012) 593–600.

[153] M. Lonnberg, J. Carlsson, Quantitative detection in the attomole range for immunochromatographic tests by means of a flatbed scanner, Anal. Biochem. 293 (2001) 224–231.

[154] G.J. Van Dam, J.H. Wichers, T.M. Falcao Ferreira, D. Ghati, A. van Amerongen, A.M. Deelder, Diagnosis of schistosomiasis by reagent strip test for detection of circulating cathodic antigen, J. Clin. Microbiol. 42 (2004) 5458–5461.

[155] E.M. Linares, L.T. Kubota, J. Michaelis, S. Thalhammer, Enhancement of the detection limit for lateral flow immunoassays: evaluation and comparison of bioconjugates, J. Immunol. Methods 375 (2012) 264–270.

[156] R. Collin, E. Schneider, L. Briggs, N. Towers, Development of immunodiagnostic field tests for the detection of the mycotoxin, sporidesmin A, Food Agric. Immunol. 10 (1998) 91–104.

[157] A. Van Amerongen, D. Van Loon, L.B.J.M. Berendsen, J.H. Withers, Quantitative computer image analysis of a human chorionic gonadotropin colloidal carbon dipstick assay, Clin. Chim. Acta 229 (1994) 67–75.

[158] P. Noguera, G.A. Posthuma-Trumpie, M. Van Tuil, F.J. Van der Wal, A. De Boer, A.P.H.A. Moers, A. van Amerongen, Carbon nanoparticles in lateral flow methods to detect genes encoding virulence factors of Shiga toxin-producing *Escherichia coli*, Anal. Bioanal. Chem. 399 (2011) 831–838.

[159] J. Bogdanovic, M. Koets, I. Sander, I. Wouters, T. Meijster, D. Heederik, A. van Amerongen, G. Doekes, Rapid detection of fungal a-amylase in the work environment with a lateral flow immunoassay, J. Allergy Clin. Immunol. 118 (2006) 1157–1163.

[160] B. Holubova-Mickova, M. Blazkova, L. Fukal, P. Rauch, Development of colloidal carbon-based immunochromatographic strip for rapid detection of carbaryl in fruit juices, Eur. Food Res. Technol. 231 (2010) 467–473.

[161] C. Suarez-Pantaleon, J. Wichers, A. Abad-Somovilla, A. van Amerongen, A. Abad-Fuentes, Development of an immunochromatographic assay based on carbon nanoparticles for the determination of the phytoregulator forchlorfenuron, Biosens. Bioelectron. 42 (2013) 170–176.

[162] M. Blazkova, P. Rauch, L. Fukal, Strip-based immunoassay for rapid detection of thiabendazole, Biosens. Bioelectron,m 25 (2010) 2122–2128.

[163] M. Blazkova, M. Koets, P. Rauch, A. van Amerongen, Development of a nucleic acid lateral flow immunoassay for simultaneous detection of *Listeria* spp. and *Listeria monocytogenes* in food, Eur. Food Res. Technol. 229 (2009) 867–874.

[164] M. Lonnberg, M. Drevin, J. Carlsson, Ultra-sensitive immunochromatographic assay for quantitative determination of erythropoietin, J. Immunol. Methods 339 (2008) 236–244.

[165] A. Abera, J.W. Choi, Quantitative lateral flow immunosensor using carbon nanotubes as label, Anal. Methods 2 (2010) 1819–1822.

[166] S.C. Lou, C. Patel, S.F. Ching, J. Gordon, One-step competitive immunochromatographic assay for semiquantitative determination of lipoprotein A in plasma, Clin. Chem. 39 (1993) 619–624.

[167] C. Liu, Q. Jia, C. Yang, R. Qiao, L. Jing, L. Wang, C. Xu, M. Gao, Lateral flow immunochromatographic assay for sensitive pesticide detection by using Fe_3O_4 nanoparticle aggregates as color reagents, Anal. Chem. 83 (2011) 6778–6784.

[168] N.M. Nor, K.A. Razak, S.C. Tan, R. Noordin, Properties of surface functionalized iron oxide nanoparticles (ferrofluid) conjugated antibody for lateral flow immunoassay application, J. Alloys Compounds 538 (2012) 100–106.

[169] W.C.W. Chan, S. Nie, Quantum dot bioconjugates for ultrasensitive nonisotopic detection, Science 281 (1998) 2016–2018.

[170] E.R. Goldman, A.R. Clapp, G.P. Anderson, H.T. Uyeda, J.M. Mauro, I.L. Medintz, H. Mattoussi, Multiplexed toxin analysis using four colors of quantum dot fluororeagents, Anal. Chem. 76 (2004) 684–688.

[171] E. Petryayeva, W. Russ Algar, I.L. Medintz, Quantum dots in bioanalysis: a review of applications across various platforms for fluorescence spectroscopy and imaging, Appl. Spectrosc. 67 (2013) 215–252.

[172] F. Esteve-Turrillas, A. Abad-Fuentes, Applications of quantum dots as probes in immunosensing of small-sized analytes, Biosens. Bioelectron. 41 (2013) 12–29.

[173] J. Li, J.J. Zhu, Quantum dots for fluorescent biosensing and bio-imaging applications, Analyst 138 (2013) 2506–2515.

[174] A.R. Clapp, I.L. Medintz, H. Mattoussi, Förster resonance energy transfer investigations using quantum-dot fluorophores, Chem. Phys. Chem. 7 (2006) 47–57.

[175] S.J. Lim, B. Chon, T. Joo, S.K. Shin, Synthesis and characterization of zinc-blende CdSe – based core/shell nanocrystals and their luminescence in water, J. Phys. Chem. C 112 (2008) 1744–1747.

[176] Z. Yu, L. Guo, H. Du, T. Krauss, J. Silcox, Shell distribution on colloidal CdSe/ZnS quantum dots, Nano Lett 5 (2005) 565.

[177] W.C. Chan, D.J. Maxwell, X. Gao, R.E. Bailey, M. Han, S. Nie, Luminescent quantum dots for multiplexed biological detection and imaging, Curr. Opin. Biotechnol. 13 (2002) 40–46.

[178] M. Bruchez, M. Moronne, P. Gin, S. Weiss, A.P. Alivisatos, Semiconductor nanocrystals as fluorescent biological labels, Science 281 (1998) 2013–2016.

[179] H. Mattoussi, J.M. Mauro, E.R. Goldman, G.P. Anderson, V.C. Sundar, F.V. Mikulec, M.G. Bawendi, Self-assembly of CdSe-ZnS quantum dot bioconjugates using an engineered recombinant protein, J. Am. Chem. Soc. 122 (2000) 12142–12150.

[180] M.F. Frasco, N. Chaniotakis, Bioconjugated quantum dots as fluorescent probes for bioanalytical applications, Anal. Bioanal. Chem. 396 (2010) 229–240.

[181] R.E. Bailey, A.M. Smith, S. Nie, Quantum dots in biology and medicine, Physica E 25 (2004) 1–12.

[182] R. Gill, M. Zayats, I. Willner, Semiconductor quantum dots for bioanalysis, Angew. Chem. Int. Ed. 47 (2008) 7602–7625.

[183] Z. Jin, N. Hildebrandt, Semiconductor quantum dots for in vitro diagnostics and cellular imaging, Trends Biotechnol. 30 (2012) 394–403.

[184] V. Biju, S. Mundayoor, R.V. Omkumar, A. Anas, M. Ishikawa, Bioconjugated quantum dots for cancer research: present status, prospects and remaining issues, Biotechnol. Adv. 28 (2010) 199–213.

[185] H. Kuang, Y. Zhao, W. Ma, L. Xu, L. Wang, C. Xu, Recent developments in analytical applications of quantum dots, Trends Anal. Chem. 30 (2011) 1620–1636.

[186] C. Frigerio, D.S.M. Ribeiro, S.S.M. Rodrigues, V.L.R.G. Abreu, J.A.C. Barbosa, J.A.V. Prior, K.L. Marques, J.L.M. Santos, Application of quantum dots as analytical tools in automated chemical analysis: a review, Anal. Chim. Acta 735 (2012) 9–22.

[187] S.G. Ding, J.X. Chen, H.Y. Jiang, J. He, W.M. Shi, W.S. Zhao, J.Z. Shen, Application of quantum dot-antibody conjugates for detection of sulfamethazine residue in chicken muscle tissue, J. Agric. Food Chem. 54 (2006) 6139–6142.

[188] J.X. Chen, X. Fei, H.Y. Jiang, Y. Hou, Q.X. Rao, P.G. Guo, S.G. Ding, A novel quantum dot-based fluoroimmunoassay method for detection of Enrofloxacin residue in chicken muscle tissue, Food Chem. 113 (2009) 1197–1201.

[189] Y.P. Chen, B.A. Ning, N. Liu, Y. Feng, Z. Liu, X.Y. Liu, Z.X. Gao, A rapid and sensitive fluoroimmunoassay based on quantum dot for the detection of chlorpyrifos residue in drinking water, J. Environ. Sci. Health B 45 (2010) 508–515.

[190] E. Tully, S. Hearty, P. Leonard, R. O'Kennedy, The development of rapid fluorescence-based immunoassays, using quantum dot-labelled antibodies for the detection of *Listeria monocytogenes* cell surface proteins, Int. J. Biol. Macromolecules 39 (2006) 127–134.

[191] C.F. Peng, Z.K. Li, Y.Y. Zhu, W. Chen, Y. Yuan, L.Q. Liu, Q.S. Li, D.G. Xu, R.R. Qiao, L. Wang, S.F. Zhu, Z.G. Jin, C.L. Xu, Simultaneous and sensitive determination of multiplex chemical residues based on multicolor quantum dot probes, Biosens. Bioelectron. 24 (2009) 3657–3662.

[192] E. Song, M. Yu, Y. Wang, W. Hu, D. Cheng, M.T. Swihart, Y. Song, Multi-color quantum dot-based fluorescence immunoassay array for simultaneous visual detection of multiple antibiotic residues in milk, Biosens. Bioelectron. 72 (2015) 320–325.

[193] X.L. Wang, G.H. Tao, Y.H. Meng, A novel CdSe/CdS quantum dot-based competitive fluoroimmunoassay for the detection of clenbuterol residue in pig urine using magnetic core/shell Fe3O4/Au nanoparticles as a solid carrier, Anal. Sci. 25 (2009) 1409–1413.

[194] L. Trapiella-Alfonso, J.M. Costa-Fernandez, R. Pereiro, A. Sanz-Medel, Development of a quantum dot-based fluorescent immunoassay for progesterone determination in bovine milk, Biosens. Bioelectron. 26 (2011) 4753–4759.

[195] N.V. Beloglazova, E.S. Speranskaya, S. De Saeger, Z. Hens, S. Abé, I.Y. Goryacheva, Quantum dot based rapid tests for zearalenone detection, Anal. Bioanal. Chem. 403 (2012) 3013–3024.

[196] S. Babu, S. Mohapatra, L. Zubkov, S. Murthy, E. Papazoglou, A PMMA microcapillary quantum dot linked immunosorbent assay (QLISA), Biosens. Bioelectron. 24 (2009) 3467–3474.

[197] H.A. Li, Z.J. Cao, Y.H. Zhang, C.W. Lau, J.Z. Lu, Combination of quantum dot fluorescence with enzyme chemiluminescence for multiplexed detection of lung cancer biomarkers, Anal. Methods 2 (2010) 1236–1242.

[198] Q. Ma, C. Wang, X.G. Su, Synthesis and application of quantum dot-tagged fluorescent microbeads, J. Nanosci. Nanotechnol. 8 (2008) 1138–1149.

[199] N.V. Beloglazova, P.S. Shmelin, I.Y. Goryacheva, S. De Saeger, Liposomes loaded with quantum dots for ultrasensitive on-site determination of aflatoxin M1 in milk products, Anal. Bioanal. Chem. 405 (2013) 7795–7802.

[200] R.H. Daniels, A.R. Watson, Immunochromatographic Methods for Detecting an Analyte in a Sample Which Employ Semiconductor Nanocrystals as Detectable Labels, US Patent Application 2002/0004246, 2002.

[201] J.L. Lambert, A.M. Fisher, Diagnostic Assays Including Multiplexed Lateral Flow Immunoassays with Quantum Dots, Patent WO/2006/071247, 2006.

[202] Z. Zou, D. Du, J. Wang, J.N. Smith, C. Timchalk, Y. Li, Y. Lin, Quantum dot-based immunochromatographic fluorescent biosensor for biomonitoring trichloropyridinol, a biomarker of exposure to chlorpyrifos, Anal Chem. 82 (2010) 5125–5133.

[203] Z. Li, Y. Wang, J. Wang, Z. Tang, J.G. Pounds, Y. Lin, Rapid and sensitive detection of protein biomarker using a portable fluorescence biosensor based on quantum dots and a lateral flow test strip, Anal Chem. 82 (2010) 7008–7014.

[204] H. Yang, D. Li, R. He, Q. Guo, K. Wang, X. Zhang, P. Huang, D. Cui, A novel quantum dots–based point of care test for syphilis, Nanoscale Res. Lett. 5 (2010) 875–881.

[205] Y. Bai, C. Tian, X. Wei, Y. Wang, D. Wang, X. Shi, A sensitive lateral flow test strip based on silica nanoparticle/CdTe quantum dot composite reporter probes, RSC Adv. 2 (2012) 1778–1781.

[206] A.N. Berlina, N.A. Taranova, A.V. Zherdev, Y.Y. Vengerov, B.B. Dzantiev, Quantum dot-based lateral flow immunoassay for detection of chloramphenicol in milk, Anal. Bioanal. Chem. 405 (2013) 4997–5000.

[207] J.G. Bruno, Application of DNA aptamers and quantum dots to lateral flow test strips for detection of foodborne pathogens with improved sensitivity versus colloidal gold, Pathogens 3 (2014) 341–355.

[208] E.A. Sapountzi, S.S. Tragoulias, D.P. Kalogianni, P.C. Ioannou, T.K. Christopoulos, Lateral flow devices for nucleic acid analysis exploiting quantum dots as reporters, Anal. Chim. Acta 864 (2015) 48–54.

[209] N.A. Taranova, A.N. Berlina, A.V. Zherdev, B.B. Dzantiev, 'Traffic light' immunochromatographic test based on multicolor quantum dots for the simultaneous detection of several antibiotics in milk, Biosens. Bioelectron. 63 (2015) 255–261.

[210] J. Hampl, M. Hall, N.A. Mufti, Y.M. Yao, D.B. MacQueen, W.H. Wright, D.E. Cooper, Up-converting phosphor reporters in immunochromatographic assays, Anal. Biochem. 288 (2001) 176–187.

[211] K. Kuningas, T. Rantanen, U. Karhunen, T. Lo1vgren, T. Soukka, Simultaneous use of time-resolved fluorescence and anti-Stokes photoluminescence in a bioaffinity assay, Anal. Chem. 77 (2005) 2826–2834.

[212] T. Soukka, T. Rantanen, K. Kuningas, Photon upconversion in homogeneous fluorescence-based bioanalytical assays, Fluoresc. Methods Appl. Spectrosc. Imaging Probes Book Ser. Ann. N. Y. Acad. Sci. 1130 (2008) 188–200.

[213] A.L. Ouellette, J.J. Li, D.E. Cooper, A.J. Ricco, G.T.A. Kovacs, Evolving point-of-care diagnostics using up-converting phosphor bioanalytical systems, Anal. Chem. 81 (2009) 3216–3221.

[214] T. Liu, L. Sun, Z. Liu, Y. Qiu, L. Shi, Rare-earth upconversion nanophosphors, Prog. Chem. 24 (2012) 304–317.

[215] T. Riuttamäki (née Rantanen), T. Soukka, Upconverting phosphor labels for bioanalytical assays, in: F.-M. Matysik (Ed.), Advances in Chemical Bioanalysis, Springer, 2014, pp. 155–204.

[216] H.J. Tanke, M. Zuiderwijk, K.C. Wiesmeijer, R.N. Breedveld, W.R. Abrams, C.J. de Dood, E.M. Tjon Kon Fat, P.L.A.M. Corstjens, The use of upconverting phosphors in point-of-care (POC) testing, in: Proceedings of SPIE 8947, Imaging, Manipulation, and Analysis of Biomolecules, Cells, and Tissues XII, 89470P, March 4, 2014.

[217] P. Corstjens, L. van Lieshout, M. Zuiderwijk, D. Kornelis, H.J. Tanke, A.M. Deelder, G.J. van Dam, Up-converting phosphor technology-based lateral flow assay for detection

of Schistosoma circulating anodic antigen in serum, J. Clin. Microbiol. 46 (2008) 171–176.
[218] R.S. Niedbala, H. Feindt, K. Kardos, T. Vail, J. Burton, B. Bielska, S. Li, D. Milunic, P. Bourdelle, R. Vallejo, Detection of analytes by immunoassay using up-converting phosphor technology, Anal. Biochem. 293 (2001) 22–30.
[219] Z. Yan, L. Zhou, Y. Zhao, J. Wang, L. Huang, K. Hu, H. Liu, H. Wang, Z. Guo, Y. Song, H. Huang, R. Yang, Rapid quantitative detection of *Yersinia pestis* by lateral-flow immunoassay and up-converting phosphor technology-based biosensor, Sens. Actuators B Chem. 119 (2006) 656–663.
[220] V.K. Mokkapati, R.S. Niedbala, K. Kardos, R.J. Perez, M. Guo, H.J. Tanke, P.L. Corstjens, Evaluation of UPlink-RSV: prototype rapid antigen test for detection of respiratory syncytial virus infection, Ann. N. Y. Acad. Sci. 1098 (2007) 476–485.
[221] P. Corstjens, M. Zuiderwijk, H.J. Tanke, J.J. van der Ploeg-van Schip, T.H. Ottenhoff, A. Geluk, A user-friendly, highly sensitive assay to detect the IFN-gamma secretion by T cells, Clin. Biochem. 41 (2008) 440–444.
[222] M. Zuiderwijk, H.J. Tanke, R.S. Niedbala, P.L. Corstjens, An amplification-free hybridization-based DNA assay to detect *Streptococcus pneumoniae* utilizing the up-converting phosphor technology, Clin. Biochem. 36 (2003) 401–403.
[223] Q. Qu, Z. Zhu, Y. Wang, Z. Zhong, J. Zhao, F. Qiao, X.Y. Du, Z. Wang, R. Yang, L. Huang, Y. Yu, L. Zhou, Z. Chen, Rapid and quantitative detection of *Brucella* by up-converting phosphor technology-based lateral-flow assay, J. Microbiol. Methods 79 (2009) 121–123.
[224] L.P. Li, L. Zhou, Y. Yu, Z. Zhu, C.Q. Lin, C.L. Lu, R. Yang, Development of up-converting phosphor technology-based lateral-flow assay for rapidly quantitative detection of hepatitis B surface antibody, Diagn. Microbiol. Infect. Dis. 63 (2009) 165–172.
[225] P. Corstjens, M. Zuiderwijk, A. Brink, S. Li, H. Feindt, R.S. Niedbala, H. Tanke, Use of up-converting phosphor reporters in lateral-flow assays to detect specific nucleic acid sequences: a rapid, sensitive DNA test to identify human papillomavirus type 16 infection, Clin. Chem. 47 (2001) 1885–1893.
[226] P. Corstjens, M. Zuiderwijk, M. Nilsson, H. Feindt, S.R. Niedbala, H.J. Tanke, Lateral-flow and up-converting phosphor reporters to detect single-stranded nucleic acids in a sandwich-hybridization assay, Anal. Biochem. 312 (2003) 191–200.
[227] P.L.A.M. Corstjens, C.J. de Dood, J.J. van der Ploeg-van Schip, K.C. Wiesmeijer, T. Riuttamäki, K.E. van Meijgaarden, J.S. Spencer, H.J. Tanke, T.H.M. Ottenhoff, A. Geluk, Lateral flow assay for simultaneous detection of cellular- and humoral immune responses, Clin. Biochem. 44 (2011) 1241–1246.
[228] P. Wang, R. Wang, W. Zhang, X. Su, H. Luo, Novel fabrication of immuno chromatographic assay based on up conversion phosphors for sensitive detection of clenbuterol, Biosens. Bioelectron. (2016). http://dx.doi.org/10.1016/j.bios.2015.10.053.
[229] H.J.M.A.A. Zijlmans, J. Bonnet, J. Burton, K. Kardos, T. Vail, R.S. Niedbala, H.J. Tanke, Detection of cell and tissue surface antigens using up-converting phosphors: a new reporter technology, Anal. Biochem. 267 (1999) 30–36.
[230] P.L.A.M. Corstjens, Z.Y. Chen, M. Zuiderwijk, H.H. Bau, W.R. Abrams, D. Malamud, R.S. Niedbala, H.J. Tanke, Rapid assay format for multiplex detection of humoral immune responses to infectious disease pathogens (HIV, HCV, and TB), Oral-Based Diagn. Book Ser. Ann. N. Y. Acad. Sci. 1098 (2007) 437–445.
[231] Y. Li, Z. Wu, Z. Liu, An immune sandwich assay of carcinoembryonic antigen based on the joint use of upconversion phosphors and magnetic beads, Analyst 140 (2015) 4083–4088.

[232] G. Glaspell, J.S. Tabb, A. Shearer, J. Wilkins, C. Smith, R. Massaro, Utilizing upconverting phosphors for the detection of TNT, Proc. SPIE 7664 (2010) 76641G.

[233] X. Song, M. Knotts, Time-resolved luminescent lateral flow assay technology, Anal. Chim. Acta 626 (2008) 186–192.

[234] X.D. Song, L. Huang, B. Wu, Bright and monodispersed phosphorescent particles and their applications for biological assays, Anal. Chem. 80 (2008) 5501–5507.

[235] T. Matsuya, S. Tashiro, N. Hoshino, N. Shibata, Y. Nagasaki, K. Kataoka, A core-shell-type fluorescent nanosphere possessing reactive poly(ethylene glycol) tethered chains on the surface for zeptomole detection of protein in time-resolved fluorometric immunoassay, Anal. Chem. 75 (2003) 6124–6132.

[236] P. Huhtinen, M. Kivela, T. Soukka, H. Tenhu, T. Lövgren, H. Härmä, Preparation, characterisation and application of europium(III) chelate-dyed polystyrene–acrylic acid nanoparticle labels, Anal. Chim. Acta 630 (2008) 211–216.

[237] Y. Xu, Q.G. Li, Multiple fluorescent labeling of silica nanoparticles with lanthanide chelates for highly sensitive time-resolved immunofluorometric assays, Clin. Chem. 53 (2007) 1503–1510.

[238] G. Rundstrom, A. Jonsson, O. Martensson, I. Mendel-Hartvig, P. Venge, Lateral flow immunoassay using europium (III) chelate microparticles and time-resolved fluorescence for eosinophils and neutrophils in whole blood, Clin. Chem. 53 (2007) 342–348.

[239] A.K. Hagan, T. Zuchner, Lanthanide-based time-resolved luminescence immunoassays, Anal. Bioanal. Chem. 400 (2011) 2847–2864.

[240] D. Tu, W. Zheng, Y. Liu, H. Zhu, X. Chen, Luminescent biodetection based on lanthanide-doped inorganic nanoprobes, Coord. Chem. Rev. 273–274 (2014) 13–29.

[241] Z. Chen, W. Zheng, H. Ping, D. Tu, S. Zhou, M. Huang, X. Chen, Lanthanide-doped luminescent nano-bioprobes for the detection of tumor markers, Nanoscale 7 (2015) 4274–4290.

[242] Q.N. Liang, J.W. Zhou, T.C. Liu, G.F. Lin, Z.N. Dong, Z.H. Chen, J.J. Chen, Y.S. Wu, Development of a time-resolved fluorescence immunoassay for herpes simplex virus type 1 and type 2 IgG antibodies, Luminescence 30 (2015) 649–654.

[243] S.M. Talha, E. Juntunen, T. Salminen, A. Sangha, T. Vuorinen, N. Khanna, K. Pettersson, All-in-one dry-reagent time-resolved immunofluorometric assay for the rapid detection of HIV-1 and -2 infections, J. Virol. Methods 226 (2015) 52–59.

[244] G.F. Lin, T.C. Liu, L.P. Zou, J.Y. Hou, Y.S. Wu, Development of a dual-label time-resolved fluoroimmunoassay for the detection of α-fetoprotein and hepatitis B virus surface antigen, Luminescence 28 (2013) 401–406.

[245] E. Sheng, H. Shi, L. Zhou, X. Hua, L. Feng, T. Yu, M. Wang, Dual-labeled time-resolved fluoroimmunoassay for simultaneous detection of clothianidin and diniconazole in agricultural samples, Food Chem. 192 (2016) 525–530.

[246] T. Le, P. Yan, J. Liu, S. Wei, Simultaneous detection of sulfamethazine and sulfaquinoxaline using a dual-label time-resolved fluorescence immunoassay, Food Addit. Contam. A 30 (2013) 1264–1269.

[247] M. Majdinasab, M. Sheikh-Zeinoddin, S. Soleimanian-Zad, P. Li, Q. Zhang, X. Li, X. Tang, J. Li, A reliable and sensitive time-resolved fluorescent immunochromatographic assay (TRFICA) for ochratoxin A in agro-products, Food Control 47 (2015) 126–134.

[248] Y. Chen, Y.M. Chi, H.M. Wen, Z.H. Lu, Sensitized luminescent terbium nanoparticles: preparation and time-resolved fluorescence assay for DNA, Anal. Chem. 79 (2007) 960–965.

[249] X.H. Xia, Y. Xu, X.L. Zhao, Q. Li, Lateral flow immunoassay using europium chelate—loaded silica nanoparticles as labels, Clin. Chem. 55 (2009) 179—182.

[250] J.Y. Hou, T.C. Liu, Z.Q. Ren, M.J. Chen, G.F. Lin, Y.S. Wu, Magnetic particle-based time-resolved fluoroimmunoassay for the simultaneous determination of α-fetoprotein and the free β-subunit of human chorionic gonadotropin, Analyst 138 (2013) 3697—3704.

[251] T.C. Liu, M.J. Chen, Z.Q. Ren, J.Y. Hou, G.F. Lin, Y.S. Wu, Development of an improved time-resolved fluoroimmunoassay for simultaneous quantification of C-peptide and insulin in human serum, Clin. Biochem. 47 (2014) 439—444.

[252] L. Du, C. Zhang, L. Wang, G. Liu, Y. Zhang, S. Wang, Ultrasensitive time-resolved microplate fluorescence immunoassay for bisphenol A using a system composed on gold nanoparticles and a europium(III)-labeled streptavidin tracer, Microchim. Acta 182 (2015) 539—545.

[253] N. Cohen, A. Mechaly, O. Mazor, M. Fisher, E. Zahavy, Rapid homogenous time-resolved fluorescence (HTRF) immunoassay for anthrax detection, J. Fluorescence 24 (2014) 795—801.

[254] I. Safarik, M. Safarikova, Use of magnetic techniques for the isolation of cells, J. Chromatogr. B 722 (1999) 33—53.

[255] Q.F. Xu, H. Xu, H. Gu, J.B. Li, Y. Wang, M. Wei, Development of lateral flow immunoassay system based on superparamagnetic nanobeads as labels for rapid quantitative detection of cardiac troponin I, Mat. Sci. Eng. C 29 (2009) 702—707.

[256] Y. Wang, H. Xu, M. Wei, H. Gu, Q. Xu, W. Zhu, Study of superparamagnetic nanoparticles as labels in the quantitative lateral flow immunoassay, Mat. Sci. Eng. C 29 (2009) 714—718.

[257] G. Zhang, Y. Liu, C. Zhang, W. Hu, W. Xu, Z. Li, S. Liang, J. Cao, Y. Wang, Aqueous immune magnetite nanoparticles for immunoassay, J. Nanopart. Res. 11 (2009) 441—448.

[258] S. Puertas, M. Moros, R. Fernandez-Pacheco, M.R. Ibarra, V. Grazu, J.M. de la Fuente, Designing novel nano-immunoassays: antibody orientation versus sensitivity, J. Phys. D: Appl. Phys. 43 (2010), 474012 (8 pp.).

[259] R.J. Flanagan, G. Martinez, Comparison of FoodChek *E. coli* test and AOAC research institute performance tested method 010603 for detection of *E. coli* O157 in raw ground beef, J. AOAC Int. 93 (2010) 922—927.

[260] R.B. Peck, J. Schweizer, B.H. Weigl, C. Somoza, J. Silver, J.W. Sellors, P.S. Lu, A magnetic immunochromatographic strip test for detection of human papillomavirus 16 E6, Clin. Chem. 52 (2006) 2170—2172.

[261] T.C. Granade, S. Workman, S.K. Wells, A.N. Holder, S.M. Owen, C.P. Pau, Rapid detection and differentiation of antibodies to HIV-1 and HIV-2 using multivalent antigens and magnetic immunochromatography testing, Clin. Vaccine Immunol. 17 (2010) 1034—1039.

Chapter 5

Rapid Multiplex Immunotests

Chapter Outline

1. INTRODUCTION

When analysing clinical and environmental samples or food products, it is usually necessary to detect a number of target analytes that may have similar or different chemical structures. Hence the development of multiparametric test systems is a logical step in the evolution of analytical methods in general and rapid immunotests in particular. This is why high attention is being paid to the development of multiparametric tests as a next step of development and substitution for monoparametric tests. The definition of multiplex testing (microarray) is a testing of multiple analytes in a single test using a single sample.

The realization of multiple detection is technologically very challenging because such tests demand full integration for ease of use and need to deliver a high analytical performance with cost-effective use of materials. This goal could be separated into two tasks. The first is to develop conditions which would be optimal for detection of each analyte at targeted level (for qualitative tests) or targeted dynamic range (for quantitative methods). In this case, sample preparation is performed before assay as preliminary step (or steps), which increases the total assay time and complexity. After sample

Comprehensive Analytical Chemistry, Vol. 72. http://dx.doi.org/10.1016/bs.coac.2016.03.009

preparation, the probe should be ready for detection of each analyte with target sensitivity, which means no additional sample dilutions or other operation. The second task is to develop methods which could incorporate sample preparation in assay setup. Microfluidic assays provide wide possibilities for this task, where the best developed multiassay setups allow to vary dilution and buffer composition for detection of different analytes (or groups of analytes) [1].

The prerequisite for multiparametric tests is that the antibodies must be highly specific. Each one must recognize only its target substance (or group of related substances in the case of group detection of, for example, related antibiotics, or aflatoxins.). As labelled specific conjugates are usually applied as cocktail, they should have negligible nonspecific interaction. Compared with assay for single analyte, multiparametric assays are usually less sensitive, partially because of the demands for establishing unified conditions for all analytes.

Multiplex assays for protein and DNA diagnostics have numerous quality of patient care and cost advantages. Several examples of multianalyte point-of-care antigen detection test systems for detection of multiple analytes are already on market, such as mariPOC (ArcDia International Oy Ltd, Turku, Finland), which can be useful to detect, for example, eight respiratory viruses (influenza A and B viruses, respiratory syncytial virus, adenovirus, human metapneumovirus and parainfluenza type 1, 2 and 3 viruses) from a single nasopharyngeal swab specimen by a fully automated, random-access immunoassay method [2].

The common approaches for multiplexing is to use multiple labels or (and) multiple test zones (Fig. 1). According to the amount of the label (or probe) multiplexing immunoassays can be classified into two dominant

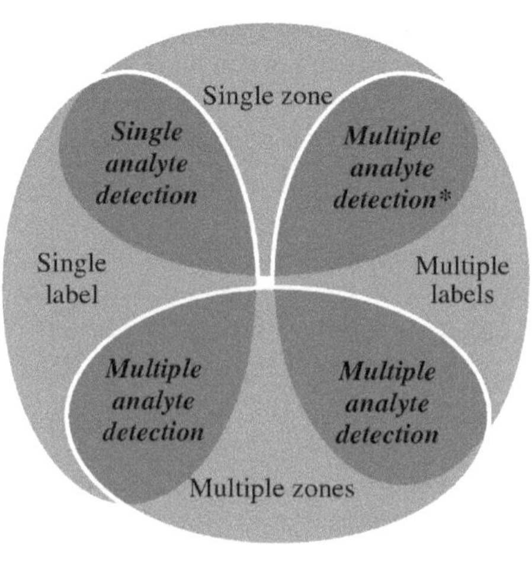

FIGURE 1 Multiplexing possibilities of immunoassay. *Limited possibilities for multiplexing.

modes. The first one is the spatial-resolved mode which utilizes a single universal probe to tag all analytes. The multiplexity is defined by test geometry. This strategy is commonly performed on ELISA plates, sensor array with fluorescent [3–5], chemiluminescent [6,7], electrochemiluminescent [8,9], colorimetric [10] or surface-enhanced Raman spectroscopy [11] detection. The second one is the multilabel mode applying multiple probes, including enzymes [12], radioisotopes [13], fluorescence dyes [14], metal ion chelates [15], nanoparticles [16] and electrochemiluminescent probes [4,5] to tag different analytes (one per analyte). The signals from different probes can be distinguished by parameters such as wavelength, mass-to-charge ratio and potential. [17]. The multiplexity is defined by the label properties, exactly the difference in these properties, which is possible to detect in the same conditions with the same equipment.

2. DETECTION OF MULTIPLE ANALYTES WITH SEPARATE TEST ZONES

2.1 Detection With Single Label

The most popular way for multiplexing is an application of separate zones with one single label. For the parallel detection of several substances, immunoreagents with different specificities bind to test zones included into one test. The classical example of such test is lateral flow immunoassay (LFIA) with multiple zones, or detection of several analytes in separate wells of single microtitre plate.

2.1.1 Lateral Flow Immunoassay

The common scheme of LFIA with multiple (two) test zones is shown in Fig. 2. A few different substances can be detected using a single test strip, as shown in studies on the simultaneous detection of two analytes, such as mytoxoxins ochratoxin A and zearalenone [18], deoxynivalenol and zearalenone [19,20]; pesticides carbofuran and triazophos [21]; biomarkers cardiac troponin I and myoglobin [22] and contaminants cyromazine and melamine [23].

For three analytes detection (pesticides imidacloprid, chlorpyrifos-methyl and isocarbophos) lateral strips with three test zones (conjugated with ovalbumin haptens) and one control zone was described [24]. Similarly, four test lines (haptens conjugated with bovine serum albumin) and one control line were located on the strip membrane for simultaneous determination of major *Fusarium* toxins, namely zearalenone, sum of T-2 and HT-2 toxins, deoxynivalenol and sum of fumonisins in wheat, oats and maize [25]. For simultaneous determination of four macrolide antibiotics (erythromycin, spiramycin, tilmicosin and tylosin) in raw milk, three antigens were immobilized as three test lines on the single-test strip [26]. In a nucleic acid LFIA, membrane technologies were used to detect simultaneously five specific

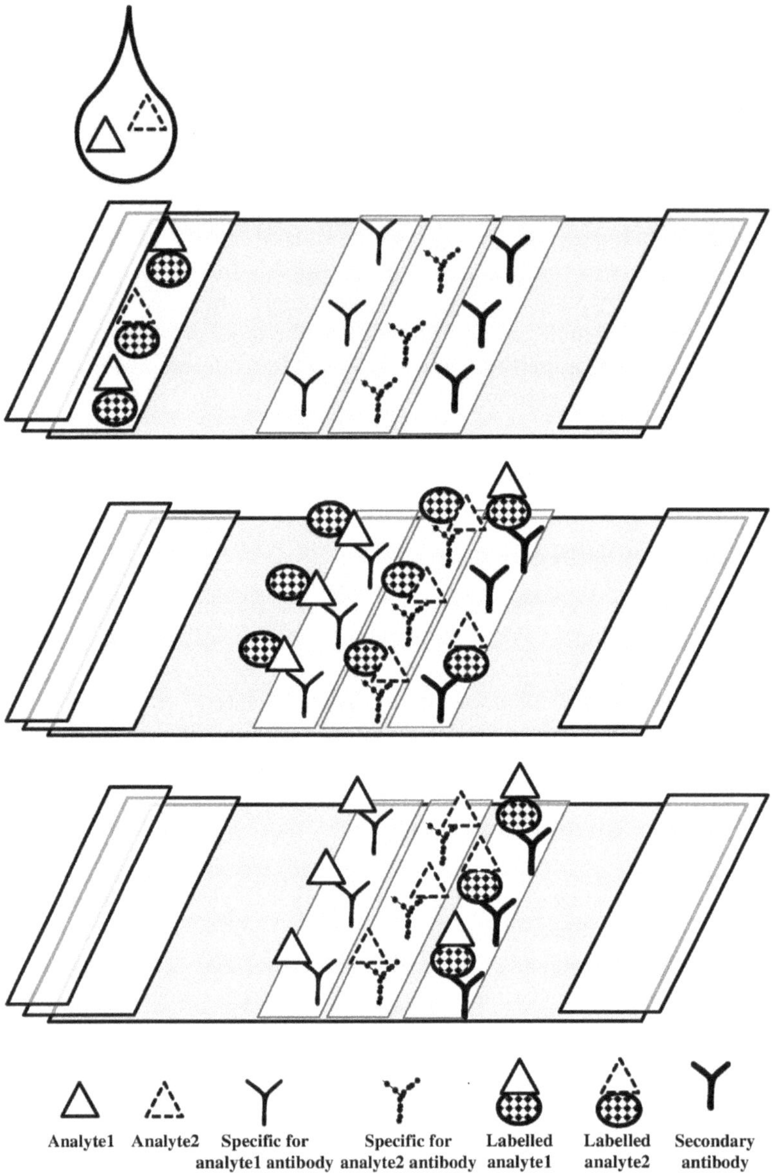

FIGURE 2 Scheme of lateral flow immunoassay with two test zones for simultaneous detection of two analytes.

nucleic-acid sequences of virulence factors of shiga toxin, produced by *Escherichia coli* [27].

The geometrical dimensions of the strip do not allow a large number of lines in the test zone, so it is practically difficult to apply more than five test lines on a single test strip and to obtain high quality analytical performance. To broaden LFIA multiassay possibilities, several test strips can be used individually or integrated into one cassette [27]. Alternatively, tests with special geometries combining different test strips have been proposed [28].

Multiplex detection can also be conducted by formulating multiplex design prototypes. Li et al. [29] fabricated a multiplex immunodisc sandwich LFIA by using AuNPs coated with a specific antibody to simultaneously and specifically detect whole-cell antigens of *Pseudomonas aeruginosa* and *Staphylococcus aureus*, with a detection range of 500–5000 CFU per mL. To optimise the multiplex design, they formulated several prototypes, such as the 'fork' and 'peace-sign', and found that the latter with a triangular centre piece can provide more effective LFIA performance than the other prototypes [30]. These approaches require a larger sample volume and more consumable materials, which increases the cost per analysis [27]. Also device and procedure could be more complicated.

A fundamental breakthrough was published by Taranova et al. [31], reporting a lateral flow microarray for the simultaneous detection of multiple analytes that combines multispot immunochip technology and immunochromatography. The test zone of the nitrocellulose support comprises a microarray spotted with up to 32 antigens that can capture AuNPs-labelled antibodies after lateral flow. The method was applied to the determination of drugs of abuse (and their metabolites) in urine, specifically of morphine, amphetamine, methamphetamine and benzoylecgonine. The detection limits (2–20 ng/mL for drugs of abuse) were comparable to those of conventional single-analyte strip methods.

Besides AuNPs, other labels were used for LFIA multisport assay. Near infrared (NIR) dyes avoid the issues of the visual range such as background from support materials and biological fluids. Compared to fluorescent dyes in the ultraviolet and visible spectrum, NIR dyes offer superb analytical sensitivity, with detection of proteins in the low pg/mL range. NIR dye was used in duplex assays to simultaneously measure interleukin-6 from 0 to 100 pg/mL (0–4.5 pmol/L) and C-reactive protein from 50 to 2500 ng/mL (0.4–20 nmol/L) on a single test strip [32].

A multiantigen printing immunoassay strip was developed to compare the antibody responses to recombinant proteins of *L. infantum* with the potential for the detection of canine visceral leishmaniasis. So for this strip the goal was not exact analytical determination, but to further characterize antibody responses to select those *L. infantum* recombinant proteins with a greater capacity to be utilized for the serodiagnosis of canine visceral leishmaniasis. Authors used set of 12 recombinant *L. infantum* antigens as test lines on the membrane of immunochromatographic strip and horseradish peroxidase (HRP) was used as label [33].

To increase LFIA multiplexing, LFIA technology could be combined with more complicated qualitative techniques. Fine example of such technique is the Lateral Flow Integrated Blood Barcode Chip (LF-IBBC), introduced by Wang et al. [34]. The LF-IBBC represents blood protein assay chip, comprised of a microfluidic layer fabricated from a hydrophilic polymer that is bonded to a glass microscope slide. Before analysis, the chip fluidic chambers were filled with the appropriate reagents (washing buffer, luminescent label and cover Ab). To operate the chip, a few microlitres of whole blood were added into last chamber, and a filter paper was inserted into the outlet slot. The capillarity of the filter paper, coupled with the hydrophilic nature of the fluidic channel surfaces, sequentially draws the blood and other reagents through the chip, with the blood drawn first; each step of a standard ELISA is automatically and sequentially executed. During the flow-through process, serum biomarker proteins are selectively captured onto specific stripes within the antibody barcodes [34].

2.1.2 Assay on Microtitre Plates and Planar Supports

A simple in situ multiplexed planar waveguide device that can simultaneously detect chloramphenicol, streptomycin and desfuroylceftiofur in raw dairy milk, without sample preparation, has been developed [35]. Samples are simply mixed with infrared dye DyLight 650 antibody prior to an aliquot being passed through the detection cartridge for 5 min before reading on a field-deployable portable instrument. Multiplexed calibration curves were produced in both buffer and raw milk. A SnapEsi reader with single use disposable cartridge, from MBio Diagnostics Inc, was used for multiplexing three different antimicrobial measurements in raw milk samples without the need for any sample preparation. The detection principle is based on the measurement of far-red fluorescence. Laser light is coupled to an injection-moulded plastic waveguide via an integrated lens moulded as part of the waveguide. This light propagates down the length of the waveguide under total internal reflection forming a uniform illumination field. At an interface between a liquid medium and the waveguide surface, an evanescent field is produced which penetrates into the liquid medium. This field decays rapidly and so, only fluorescent dyes held in close proximity to the surface will undergo excitation [36]. Immunochemical reagents were accurately immobilized to the surface using nano-spotting technology, thus providing multiplexing capability, while their binding partners were fluorescently labelled and used within the liquid medium [35].

As solid support, agarose-modified glass slides with several physically isolated subarrays were used for detection of six mycotoxins (aflatoxin B1, aflatoxin M1, deoxynivalenol, ochratoxin A, T-2 toxin and zearalenone) with cyanine dye Cy3 as a label [5].

Microtitre plate–based format provides wide possibilities for development of several immunoreagent zones for multiplexing. The simplest way to obtain multiplexing is a combination of ELISA with array analysis. This way, immunoreagents for each analyte are immobilized in separate wells (lines).

A lot of assays have been developed with HRP label. Rapid ELISA for simultaneous detection of multiple analytes presents typical example of detection in separate zones with single label. A variant of the ELISA with kinetic regime of signal evaluation allows Urusov et al. [37] to reduce assay time of simultaneous detection of three mycotoxins (aflatoxin B1, ochratoxin A and zearalenone) to 25 min. The use of biotin with an extended spacer together with a streptavidin–polyperoxidase conjugate provided high signal levels, despite that these interactions occurred under nonequilibrium conditions. The proposed kinetic ELISA was accomplished with the use of several additional reagents and steps, thereby enabling the effective incorporation of the label in the complexes to be detected:

1. The biotin–streptavidin interaction was exploited for the detection of primary antibodies, as this complex has a higher binding constant (10^{15} M^{-1}) compared with the interaction between primary antibodies and antispecies antibodies.
2. Specific primary antibodies were modified with a biotin ester containing a 14-atom spacer.
3. The streptavidin moiety was modified with a polymer that enabled the coupling of several peroxidase molecules [37].

As an alternative for enzyme, DNAzyme-based ELISA, termed DLISA, was developed as an enzyme-free method, combining a catalytic and molecular beacon system with a cation exchange reaction for ultrasensitive multiplex fluorescent immunosorbent assay [38].

Different modifications of microtitre plates and related supports have been established. To achieve the multiplexed detection of four pathogens (*E. coli* O157:H7, *Yersinia enterocolitica*, *Salmonella typhimurium* and *Listeria monocytogenes*), a polystyrene 96-well microtitre plate format has been designed, in which each main well contains four subwells with separately immobilized antibodies specific for each bacteria. After sample applications, a mixture of peroxidase-labelled polyclonal antibodies against the four bacteria was added and the peroxidase activity of the bound polyclonal labelled antibodies in each well was measured by an enhanced luminol-based chemiluminescent cocktail [39].

A specially designed 16-well slide containing a miniaturized array of test spots was developed for human prostate-specific antigen together with low cost optical device for reading colorimetric and fluorescent immunodiagnostic test results. Arrays are illuminated with either LEDs or lasers, while transmitted or emitted light is captured through a long-pass filter, allowing two different types of optical measurement to be performed within the same device. This device was used to read results from an array of antibodies conjugated with either an enzymatic or fluorescent tag resulting in a coloured or fluorescent readout [40].

Multiplex antibody microarray method to simultaneously capture and detect *E. coli* O157:H7 and *Salmonella enterica* serovar *typhimurium* (*S. typhimurium*), as well as a biomolecule (chicken immunoglobulin G or IgG employed as a proteinaceous toxin analogue) in a single sample was developed using 96-well microplate as solid support. Microarrayed spots of captured antibodies against the targeted analytes were printed within individual wells of 96-multiwell microtitre plates, and a sandwich assay with fluorescein- or Cy3-labelled reporter antibodies was used for detection. Multiplex detection of the two bacteria and the IgG in buffer and in culture-enriched ground beef filtrate was established with a total assay (including detection) time of c. 2.5 h [41].

Original multiassay design with separate spots immobilized within the nitrocellulose/cellulose acetate membrane of a 96-well filtering microtitre plate bottom was developed to detect auto-antibodies directly in patient sera. Microarrays with 12 different probes were used for detection of captured auto-antibodies using a staining approach based on alkaline phosphatase labelling. In real samples, nine proteins from the 12 were shown to generate specific signal and five antigens were shown to have interaction with more than 10% of the positive sera from cancer patients [42].

2.1.3 Flow-Through Tests

This group of methods includes two kinds of setup geometries and thus tests performances. The first one is membrane-based flow-through methods. One or more test zones as spots of specific antibodies and optionally control zone as spot of antienzyme antibody are placed on the surface of 2D porous support. As a label, HRP [43,44] was commonly used, but Burmistrova et al. [45] found the alkaline-conjugates to be less prone to matrix interferences.

Two variants of zone location were described.

1. Location of each zone (including the control zone) into separate well [46] (Fig. 3A) on a single membrane with parallel application of portions of single sample into each well. Test device for simultaneous multitoxin analysis (deoxynivalenol and other trichotecenes) included membrane coated with different antitoxin antibodies fixed to each well, and below the membrane a cellulose layer support was attached as a filter pad. Test sequence for individual wells was performed in the following order: sample solution, enzyme-labelled antigen solution and chromogenic substrate solution. Excess liquid is absorbed by a filter pad below the membrane. In the presence of analyte, colour development is suppressed, while negative samples give maximum colour development.
2. Location of several zones side by side in single membrane area with application of one sample portion to interact with all zones (Fig. 3B) [43,45]. This variant was developed for detection of mycotoxins in food and feed, namely, aflatoxin B1, zearalenone, deoxynivalenol, ochratoxin A and fumonisin B1 in cereal samples [47]; ochratoxin A, zearalenone

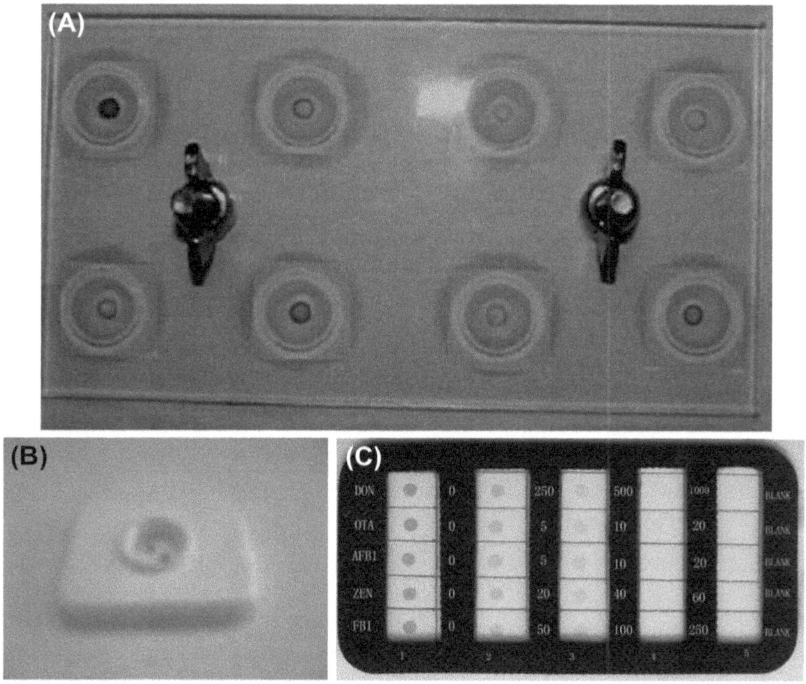

FIGURE 3 (A) Prototype of an 8-well immunofiltration test device for simultaneous multitoxin analysis *(Reproduced from E. Schneider, V. Curtui, C. Seidler, R. Dietrich, E. Usleber, E. Martlbauer. Rapid methods for deoxynivalenol and other trichothecenes, Toxicol. Lett. 153 (2004) 113–121.)*. (B) Test with several zones side by side in single membrane area. (C) Test for detection of five toxins at five different levels *(Reproduced from Q.H. He, Y. Xu, D. Wang, M. Kang, Z.B. Huang, Y.P. Li, Simultaneous multiresidue determination of mycotoxins in cereal samples by polyvinylidene fluoride membrane based dot immunoassay. Food Chem. 134 (2012) 507–512.)*.

and fumonisin B1 in cereal grains and silage [45]; ochratoxin A, aflatoxin B1, deoxynivalenol and zearalenone in peanut cake, maize and cassava flour [44]; aflatoxin B1 and ochratoxin A in chili samples [43], and also for detection of the pesticides, carbaryl and endosulphan [48]. This variant was adapted also to detect analytes at different concentration levels (Fig. 4) [45].

As an intermediate variant, He et al. [47] developed test for detection of five mycotoxins (aflatoxin B1, zearalenone, deoxynivalenol, ochratoxin A and fumonisin B1) at five different levels. To circumvent the problem of cross-contamination and to facilitate the rapid screening of a batch of samples in a single test device, reaction zones were separated by rubber fence and scotch tape (Fig. 3C).

It is important to mention that this assay format is very prone to matrix influence, because application of liquid sample takes place immediately on the

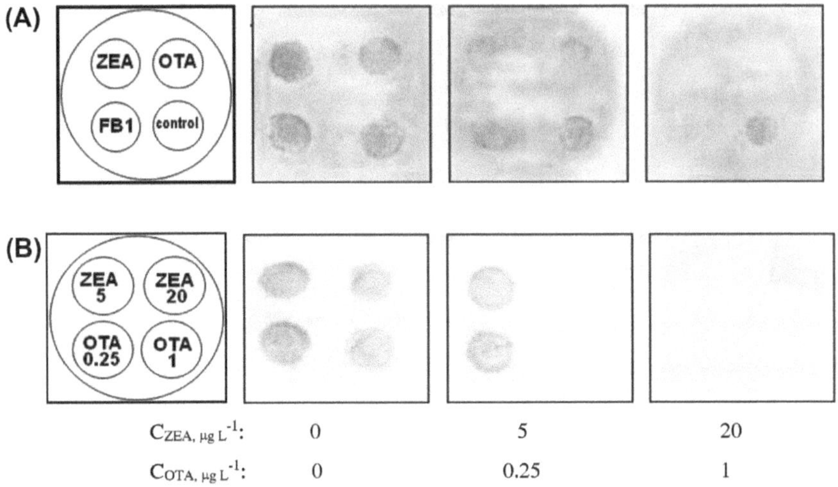

FIGURE 4 (A) Membrane containing three detection test spots (antianalyte antibody coated) and control spot (antialkaline phosphatase antibody coated) and membrane images for multiple analytes detection. (B) Membranes for two analytes detection at two cut-off levels.

immunoreagent spots. This means all impurities presented in sample could influence the immunoassay outcomes. The limitation of the volume of washing buffers, related with restricted volume of absorbent, additionally complicates the situation. The problem could be partially solved by the application of sample purification, for example with solid phase clean up [49], or by changing absorbent pads during assay procedure, which was not fixed to the reaction membrane [50]. Between steps, it was possible to wash with a stream of washing buffer from a wash bottle [51].

As alternative for binding of specific antibodies to the test zones, format with bounded analyte protein conjugates was used for detection of five mycotoxins [47]. This allows to use commercial labelled with enzyme secondary antibody and analytical signal source.

The second group of flow-through tests, developed for multiassay, is multiplex flow-through microcolumns. One or more test layers with bound specific antibodies and optionally control layer with bound antibody specific to the label (or it component) are placed and fixed inside the transparent plastic column. To prepare solid support for test and control layers, specific antibodies were attached to the sepharose gel or polyethelene porous frits. These tests allow scientists to use different variants of microcolumn setups with different amount of test layers and possibilities to use separate clean-up columns for sample purification. A variant of microcolumn setup for simultaneous detection of two analytes: separate clean-up column and flow-through detection column are presented in Fig. 5.

As labels, enzymes HRP [52–54] or liposomes loaded with fluorescence QDs [54] were used. An application of microcolumn multiparametric tests

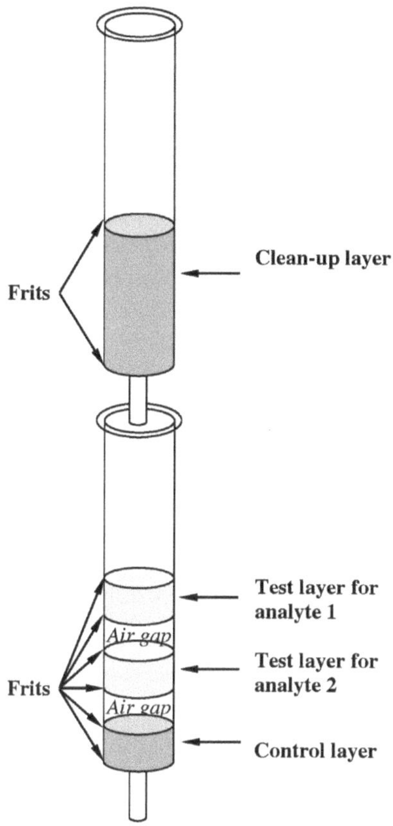

FIGURE 5 A variant of microcolumn setup for simultaneous detection of two analytes: separate clean-up column and flow-through detection column are presented.

with peroxidase label was described for detection of mycotoxins aflatoxin B1 and ochratoxin A in spices [52]; 2,4,6-trichlorophenol and ochratoxin A in red wine [53]; groups of sulphonamides and quinolones in milk [54]; mycotoxins ochratoxin A, fumonisin B1, deoxynivalenol and zearalenone detection at cut-off values of 3, 1250, 1000 and 200 μg/kg, respectively, in peanut cake, maize and cassava flour [44]. Evaluation of results could be performed visually or using multilayer reader from Senova immunoassay systems (Germany). Multiplex column assay with liposomes loaded with fluorescence QD labels was developed for groups of sulphonamides and quinolones in milk [54], and qualitative LODs were as low as 1 and 0.5 ng/mL for the sulphonamides and quinolones, respectively, which is approximately a 10-fold improvement compared to the use of HRP as a label. For multiplex determination of four mycotoxins in cereals, the achieved QD label cut-off values were 500, 100, 2 and 100 mg/kg for deoxynivalenol, zearalenone, aflatoxin B1 and T2-toxin, respectively. For simplification of multiassay results' evaluation, the conjugates with QDs of different colours were used [55].

2.1.4 Microfluidic Multiplex Immunoassays

This group includes more advanced settings and methods, but progress in various fields of science and technology makes lab-on-chip technologies closer to demands for rapid tests in the terms of simplicity of application, but provided with better analytical features and multiplexity, robustness, low-cost and ease of use. A large number of microfluidic multiplex immunoassays have been developed, many targeted at low-resource settings [55–57]. 'Lab-on-a-chip' have garnered a great deal of attention because of the possibility of miniaturization and automation. The most popular target for these research is blood multiassay, especially microfluidic systems providing complete sample-to-result process [1,58–60].

However, examples of fully integrated systems including a sample preparation step, especially with the capabilities of handling real samples such as whole blood, have been rare [61]. It is mainly because of the complexity and the cost issues in the fluidic control of multiple reagents.

Multianalyte detection from whole blood has been achieved for antibody detection [61], for differentiating between infections that cause similar clinical symptoms [1]. Multianalyte all-in-one microfluidic detection presents special challenges, because the preferred diagnostic targets for each pathogen often require different sample preparation steps and assay formats that may be difficult to implement in a single device [1].

Interesting kind of platform is 'lab-on-a-disc'. A range of biological assays such as immunoassays [62], cell lysis and homogenization [63] and DNA analysis [64] have been demonstrated on a centrifugal microfluidic platform. A fully integrated portable device that can perform not only immunoassay but also multiple kinds of biochemical analysis simultaneously on a disc starting from whole blood has been reported by Lee et al. [60]. Microchip for whole blood assay allows to separate plasma and to use it for assay both undiluted and diluted, when it is necessary [1].

Not only new constructions, but also new materials sensitive to physical (or chemical) conditions make progress in chip development. The use of a novel photosensitive material in a 'one-step' fabrication process allowed authors [65] the rapid fabrication of microfluidic components and interconnection port simultaneously for multiantibiotic competitive immunoassay. Prefilled microfluidic cartridges were used as binary response rapid tests for the simultaneous detection of three antibiotic families (sulphonamides, fluoroquinolones and tetracyclines) in raw milk.

2.2 Detection With Multiple Labels

Additional information about different contaminants could be obtained by using multicoloured labels for binding in different bands of the test zones [27]. It also could make evaluation of results easier. The presence or the absence of binding in each of the zones can be monitored visually or by a portable detector.

An interesting 'traffic light' LFIA was developed for the simultaneous detection of several compounds. The system comprises three lines of different colours on a test strip, related to three different test zones for detection of three different analytes and visualized with QDs of three different luminescence colours (luminescence maximum wavelengths at 525, 585 and 625 nm; Fig. 6). Test allowed to identify an analyte of interest based on the visible colour of the line formed (qualitative analysis), and to determine the amount of the analytes present based on the luminescence intensity of the lines (quantitative analysis). The test system exhibited high sensitivity, with limits of detection for ofloxacin, chloramphenicol and streptomycin of 0.3, 0.12 and 0.2 ng/mL, respectively. These values are 80–200 times lower than those achievable with traditional ELISA using the same antibodies. Using the 'traffic light' assay, antibiotics could be detected in milk samples within 10 min without any sample preparation [66].

Multicolour QD-based immunofluorescence assay were developed on microtitre plate for quantitative and visual detection of another set of antibiotics (streptomycin, tetracycline and penicillin G) in milk. Specific antibodies were conjugated to QDs with different emission wavelengths (QD520 nm, QD565 nm and QD610 nm) to serve as detection probes [67]. Additionally

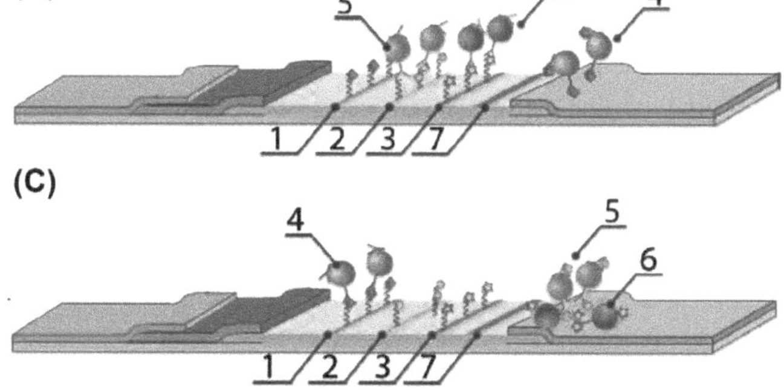

FIGURE 6 Principle of the competitive detection of three analytes using 'traffic light' immunochromatographic test: 1 – test zone for analyte 1 (streptomycin); 2 – test zone for analyte 2 (chloramphenicol); 3 – test zone for analyte 3 (ofloxacin); 4-6 – labelled with QD antibody, specific for analytes 1-3; 7 – control line). (A) Test strip before the assay, (B) assay results for the sample containing streptomycin, (C) assay results for the sample containing chloramphenicol and ofloxacin. *The picture from B.B. Dzantiev, N.A. Byzova, A.E. Urusov, A.V. Zherdev, Immunochromatographic methods in food analysis, Trends Anal. Chem. 55 (2014) 81–93.*

with quantitative detection using luminescence microplate reader, visual detection was also suggested. The QD520 nm, QD565 nm and QD610 nm could present hunter green, chartreuse and vermillion fluorescence colours, respectively, under excitation with blue light. Based on this, the three different antibiotics could be determined by different fluorescence colours of QDs, and the quantities of each kind of antibiotic could be determined based on the different shades of each colour image (using provided colour cards). Similar microtitre plate format with multicolour QDs as label was used for simultaneous quantitative detection of five mycotoxins (deoxynivalenol, zearalenone, aflatoxin B1, T2-toxin and fumonisin B1) in one plate [68].

Microcolumn-based format has also been used for developing of the tests with multicolour QDs for detection of four mycotoxins in cereals and the achieved QD labels cut-off values were 500, 100, 2 and 100 mg/kg for deoxynivalenol, zearalenone, aflatoxin B1 and T2-toxin, respectively [68]. Similar approach (but as label, liposomes multiloaded with quantum dots were used) for groups of sulphonamides and quinolones [54] was far more sensitive than HRP-based immunoassays. Its qualitative limits of detection were as low as 1 and 0.5 ng/mL for the sulphonamides and quinolones, respectively, which is approximately a 10-fold improvement com-pared to the use of HRP as a label.

3. DETECTION OF MULTIPLE ANALYTES WITHIN SINGLE TEST ZONE

For using one test zone for detection of several analytes, it is necessary to use labels that could provide separable signals in the same conditions. For common analytical methods, different approaches such as chemometric could be used for signal separation, but for rapid tests and test techniques in general, it is more convenient to have signals separated without additional efforts. Different kinds of signals, including optical and electrochemical, could be used for multiplex assay in single test zone.

Among optical signals, the simplest one is colorimetric instrumental or visual detection, which is used for evaluation of presence (concentration) of two or more analytes within single test zone. As a fine example, multiplexed dot immunoassay with multicolour nanoparticles was reported by Panfilova et al. [69]. Ag nanocubes, Au/Ag alloy nanoparticles and Au/Ag nanocages were prepared by a galvanic replacement reaction between the Ag atoms of silver nanocubes and Au ions of tetrachloroauric acid. Depending on the Ag/Au ratio, the particle plasmon resonance was tuned from 450 to 700 nm and the suspension colour changed from yellow to blue. The particles of yellow, red and blue suspensions were functionalized with chicken, rat and mouse immunogamma globulin (IgG) molecular probes, respectively. The multiplex capability of the assay was illustrated by a proof-of-concept experiment involving simultaneous one-step determination of target molecules (rabbit antichicken, antirat and antimouse antibodies) with a mixture of fabricated conjugates. Under naked eye examination, no cross-coloured spots or nonspecific bioconjugate adsorption were observed, and the low detection limit was about 20 fmol [69].

Luminescence provides more possibilities because of both variability of signal nature (spectral or time resolution) and different mechanisms of signal generation.

An application of luminophores with excitation by single light source and luminescence in different spectral areas is the most suitable for multiplex rapid tests. QDs' wide absorbance spectra and narrow luminescence peaks make them very attractive signal source for such application. LFIA for simultaneous detection of two analytes (cardiac markers alpha fetoprotein and carcinoembryonic antigen) with specific antibodies for two analytes attached into single test line was proposed by Wang et al. [70]; scheme is presented in Fig. 7. After dropping of solution to analyse onto the sample pad, it migrates towards the other end of the strip carrying along antialpha fetoprotein antibody conjugated with QDs λ max = 546 nm and anticarcinoembryonic antigen antibody conjugated with QDs λ max = 620 nm. To evaluate analytes' concentration, the luminescence intensity of test line and control line on a single strip were both measured and the ratio of the signals was used to offset the background factors.

Similar LFIA format with luminescent label used yellow green and dark red fluorescent polymer microspheres for quantifying chronic wound biomarkers, interleukin 6 and tumour necrosis factor, using strip with single test line. It was found that two kinds of microspheres were completely spectrophotometrically distinguishable from each other on the ESE Quant Lateral flow reader [71].

Similar approach was published for microtitre plate format [68]. For the double-analyte multiplex, two different specific antibodies were co-immobilized in one single well (Fig. 8). Mycotoxins zearalenone and aflatoxin B1 were simultaneously determined, provided their conjugates were labelled with QDs which are luminescent in different parts of the spectrum, by scanning the assay outcome at two different wavelengths using microtitre plate reader. The same format was used with silanized liposomes loaded with luminescent QDs (orange and green). Using this label, allowed to increase sensitivity: the IC50 for zearalenone was four times lower than for the QD-based assay (0.17 ng/L vs 0.71 ng/L). For AfB1, the IC50 were decreased five times (0.09 ng/L vs 0.55 ng/L) [72].

Interestingly, some labels and techniques give possibility for both spectral and time resolution. Lanthanide ions have luminescence spectra with well-defined peaks for each ion, which could be used for spectral selection of lanthanide label signals from the single test zone. But the technique based on multiple lanthanides with distinguishable lifetime as the probes demands resolution of the analytical signals from different analytes in separate time windows typically employs [73–75] a highly sensitive direct dual-labelled

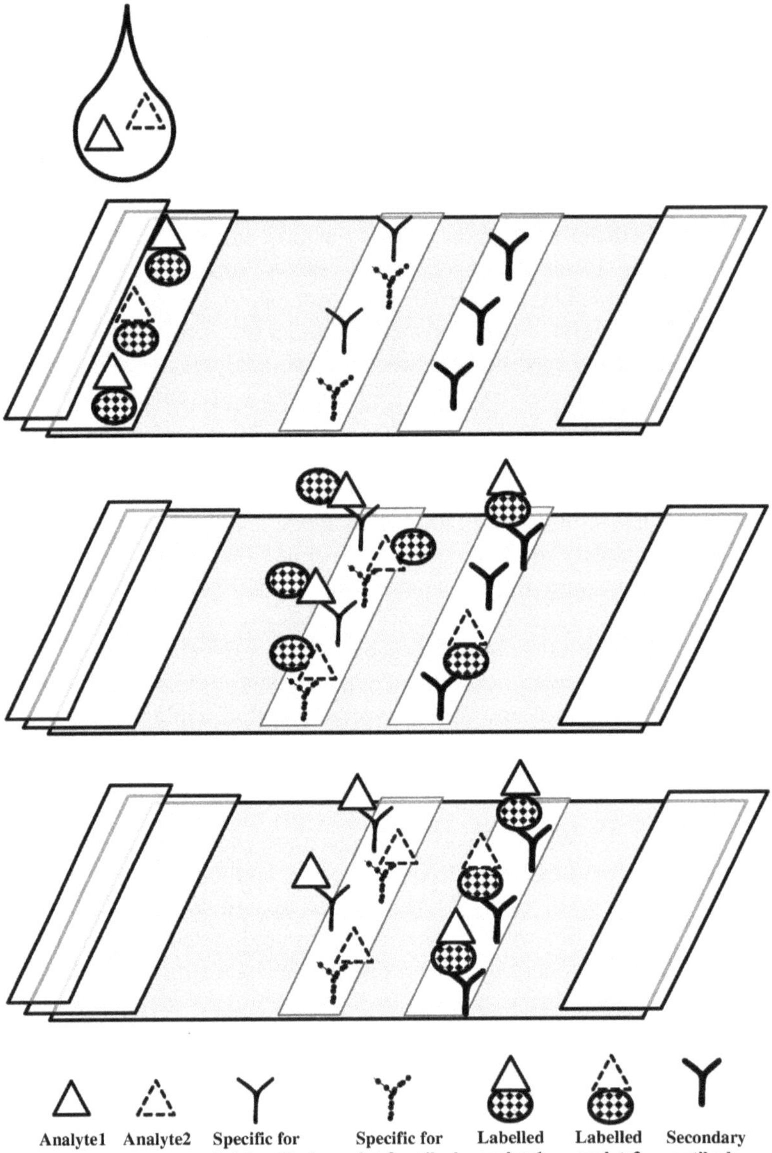

FIGURE 7 Scheme of lateral flow immunoassay with single test zone for simultaneous detection of two analytes.

time-resolved fluoroimmunoassay to detect parathion and imidacloprid in food and environmental matrices [73], clothianidin and diniconazole in food samples [74], human serum thyroid-stimulating hormone and thyroxin [75]. Europium (Eu^{3+}) and samarium (Sm^{3+}) were used as fluorescent labels by

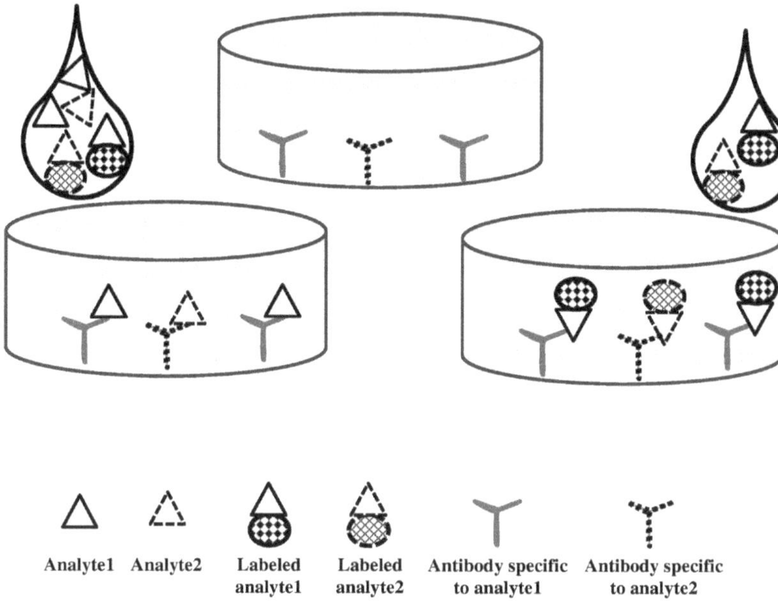

FIGURE 8 Multiplex Competitive ELISA for detection of two analytes in single well. Left: presence of analytes; right: absence of analytes.

coupling separately with specific to different analytes antibody, thus to resolve the signals from different analytes in different time windows. Time-resolution phenomenon has also been investigated and utilized in chemiluminescence analysis [17].

Chemiluminescence gives signals which are possible to resolve both spectrally and temporally. For simultaneous detection in one sport immunocomplexes formation, labelled with two enzymes: alkaline phosphatase and HRP, time-resolved strategy was described by Han et al. [17]. The strategy was performed based on the distinction of the kinetic characteristics of different chemiluminescence reaction systems, which allowed detection of multiple analytes in different time windows. HRP and alkaline phosphatase were adopted as the signal probes to tag the two antigens due to their very different chemiluminescence kinetic characteristics. After the competitive immunoreactions, the two chemiluminescence signals were simultaneously triggered by adding the chemiluminescence co-reactants. Then the signals for analytes (clenbuterol and ractopamine) were in turn detected after 0.6 s and 25 min of the reaction triggering. Due to the distinguishable detection time windows for HRP and alkaline phosphatase, the cross-talk resulting from the mixed chemiluminescence reaction systems was effectively avoided [17].

Another interesting principle to obtain signals separately from alkaline phosphatase and HRP in one spot was suggested by Chong et al. [76]. In the proposed assay fluorescein was formed from the reaction of fluorescein diphosphate and immunocomplex conjugated with alkaline phosphatase. Resorufin was formed from the reaction between Amplex Red and H_2O_2 in the presence of immunocomplex conjugated with HRP. When 1,1′-oxalyldiimidazole chemiluminescence reagents (H_2O_2 in isopropyl alcohol, 1,1′-oxalyldiimidazole in ethyl acetate) were injected in a test tube or strip-well containing fluorescein and resorufin formed from above two reactions, a chemiluminescence emission spectrum having two peaks (518 nm for fluorescein and 602 nm for resorufin) was observed. The peaks can be independently quantified with an appropriate statistical tool capable of deconvoluting multiple emission peaks. But the principle opens perspectives for application as rapid tests also, if signals could be fixed and processed with simple reader.

A sandwich-format electrochemical immunosensor for simultaneous determination of three cancer biomarkers using the carbon–gold nanocomposite as immunoprobes was introduced. The nanocomposite was used to fabricate three electrochemical immunoprobes through adsorbing thionin, 2,3-diaminophenazine and Cd^{2+}, and fixing three kind of antibodies, respectively. The ionic liquid reduced graphene oxide with superior conductivity was used as biosensor platforms for immobilization of proteins. The AuNPs decorated on the nanocomposite provided extra binding sites for the three antibodies for detection of three tumour markers – carcinoembryonic antigen, prostatespecific antigen and α-fetoprotein, respectively. Three separate signals can be detected directly in a single run through square wave voltammetry. The voltammetric peaks near −0.05 V, −0.35 V and −0.65 V comes from the thionin, 2,3-diaminophenazine and Cd^{2+}, respectively, which represent the existence of carcinoembryonic antigen, prostatespecific antigen and α-fetoprotein, respectively. The SWV peak currents of the thionin, 2,3-diaminophenazine and Cd^{2+} increased with the increment of concentrations of tumour markers in solutions [77].

4. MICROCARRIER-BASED MULTIPLEX TECHNOLOGIES

For multiplex immunoassay realization, it is necessary to distinguish different binding events in parallel. The most common and versatile approach is the planar array, in which the probe molecules are immobilized on a substrate and encoded by the coordinate of their positions (Section 2.1). In alternative approach, the reactions are carried out on individual microcarriers – each carrier having a particular recognition molecule bound to its surface (Fig. 9). This method allows uniquely encoded microcarriers to be mixed and subjected to an assay simultaneously. Those microcarriers that show a favourable reaction between the attached molecule and the target analyte – which can be registered, for example, by a resulting fluorescent signal – can then have their codes read, thereby revealing the identity of the molecule that produced the favourable reaction [78]. As encoded elements for suspension arrays,

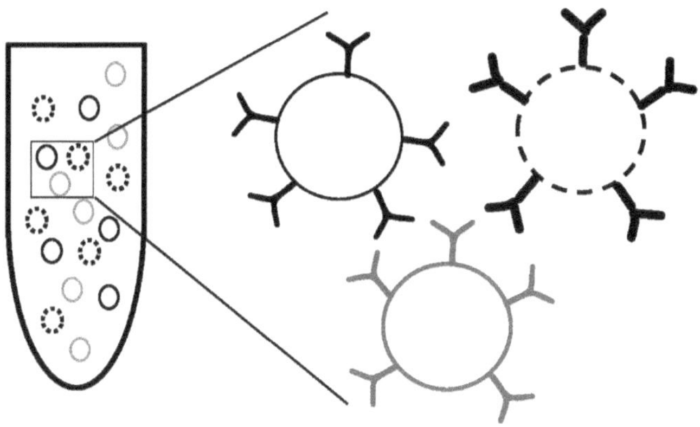

FIGURE 9 A suspension array composed of recognition molecules attached to encoded particles.

fluorescent molecules; vibrational signatures; quantum dots; discrete metallic layers; photonic crystals; shape or radio frequency have been described in scientific literature. In comparison with the planar arrays, suspension arrays show high flexibility, fast reaction, and good repeatability for detecting. In addition, suspension arrays consume less analyte sample and cost. Among a number of different suspension arrays, the spectrum-encoded suspension arrays are well used due to their simplicity in both encoding and detection [79]. Fluorescent dyes and quantum dots are the main spectrum-encoding elements, and the beads encoded by fluorescence have been applied to genotyping, proteomics and diagnostics by Luminex (Austin TX), Becton Dickinson Biosciences (San Jose, CA) and DiaSorin (Saluggia, Italy).

Most of the practically used suspension arrays are composed of polymer microspheres internally doped with one or more fluorescent dyes. Polystyrene microspheres become swollen when suspended in an organic solvent, allowing the dye molecules to diffuse into them, but when the microspheres are transferred to an aqueous solution they shrink and the dye molecules become entrapped. By trapping dyes with different emission spectra at different concentrations (and thus intensities), microspheres with unique spectral codes are obtained. For multiplexed detection, a reporter dye is required and the region of the spectrum that is occupied by its emission profile is not available for encoding. Luminex supplies microspheres that are encoded with organic dyes as part of their xMAP liquid array technology. They encode 5.5-μm microspheres with two dyes at 10 different concentrations to produce up to 100 different sets of microspheres. Each set is matched to an own probe molecule that confers specificity on the microspheres in multiplexed assays [80].

One of the most popular microcarrier-based technologies with wide possibilities for multiplexing is flow cytometry – a technology that measures intrinsic and evoked optical signals from single cells or cell-sized particles in a moving fluid stream. The idea of using flow cytometers to perform immunoassays dates from the late 1970s [81]. Following developments have given this technology the potential to evolve into an efficient, relatively inexpensive and flexible tool for both researchers and clinicians. In this technology, a fluidics system takes a suspension of single microbeads and passes them single file through the flow cell where the particles produce optical signals, which are either an intrinsic property of the analyte or consist of a chromophore tagged to an analyte. The combination of conventional immunoassays with microbeads led to a new wave of instrumentation that covers the best of both worlds of enzyme-linked immunoassays and flow cytometry. In the past 30 years, suspension arrays have become the gold standard for immunophenotyping, including clinical diagnosis of AIDS (enumerating T-cell subsets) and certain cancers [82]. A good critical review devoted to multiplexed microbead immunoassays by flow cytometry was published by Krishhan et al. [83], and Wilson et al. [80].

The most popular label for the microbead array is fluorescent; for excitation, laser with the wavelength different from the laser wavelength for microcarriers identification is used. Other variants of detection are the subject of development now. Multiplex chemiluminescent mycotoxin immunoassay suspension array system was developed by combining the silica photonic crystal microspheres-encoding technique and a chemiluminescent immunoassay method. The microspheres were used as a carrier and encoded by their reflectance peak positions, which overcome the potential interference with detection luminescence. Aflatoxin B1, fumonisin B1 and ochratoxin A artificial antigens were immobilized on the surfaces of microspheres. Horseradish peroxidase was used as a labelling enzyme in the enzyme-catalysed chemiluminescence system [84].

Microcarrier-based multiplex technologies was integrated with biofunctionalized QDs to develop a competitive microbead immunoassay using luminescent QD labels for simultaneous detection of two analytes, the cyanobacterial toxin microcystin-LR and the polycyclic aromatic hydrocarbon compound benzo[a]pyrene. The assay was carried out in two steps: the competitive reaction of multiple targets using their exclusive sensing elements of QD/antibody detection probes and antigen-coated microsphere, and the subsequent flow cytometric analysis. The fluorescence of the QD-encoded microsphere was thus found to be inversely proportional to target analyte concentration. Under optimized conditions, the proposed assay performed well within 30 min for the identification and quantitative analysis of the two analytes [85].

Magnetic carriers provide additional possibilities for operating and preconcentration. Magnetic nano- and microparticles are widely used in modern assay because of the possibility to integrate magnetic beads modified with immunoreagents (usually antibody) for immunoseparation. Magnetic field allows easy operation with movement and location of magnetic beads to reach very high preconcentration factors. These provided magnetic particles application for multiplex assay.

A fine example of an optomagnetic immunoassay technology is based on nanoparticles that are magnetically actuated and optically detected in a stationary sample fluid. The dynamic control of nanoparticles by magnetic field impacts the key immunoassay process steps, giving speed, assay control and seamless integration of the total test. The optomagnetic technology enables high-sensitivity one-step assays of drugs of abuse and cardiac marker troponin I in blood serum/plasma and whole saliva. At the first step, the reaction microchamber was filled with liquid sample, which redispersed magnetic nanoparticles covered with specific antibodies. In the presence of analyte, nanoparticles captured analyte in solution and bound to the sensing surface covered with antibody through sandwich immunocomplex formation. Free and weakly bound nanoparticles were removed from the sensing surface by magnetic forces. Light reflects from the sensor surface with an intensity that depends on the concentration of nanoparticles at the sensor surface, by the mechanism of frustrated total internal reflection [86].

Magnetic nanobeads were used for immunoseparation with QDs as fluorescent labels for rapid, sensitive and simultaneous detection of three major pathogenic bacteria, *S. typhimurium*, *E. coli* O157:H7 and *L. monocytogenes*, in food products. Both streptavidin-conjugated magnetic nanobeads and QDs (530-, 580-, and 620-nm emission wavelength) were separately coated with biotinylated anti-*Salmonella*, anti-*E. coli* and anti-*Listeria* antibodies. First, magnetic nanobeads were mixed with a food sample to capture the three target bacteria. After being magnetically separated from the sample, the magnetic nanobeads-cell conjugates were mixed with the QDs to form the triple immunocomplexes, and unattached QDs were removed. The fluorescence intensity of the immunocomplexes was measured at wavelengths of 530, 580 and 620 nm to determine the populations of *S. typhimurium*, *E. coli* O157:H7 and *L. monocytogenes*, respectively [87].

5. HOMOGENEOUS MULTIPLEX ASSAY

Contrary to the methods described earlier, fluorescence polarization immunoassay (FPIA) is a homogeneous method; it functions without attachment of immunoreagents to solid surfaces. In this format, the analyte is labelled with a fluorophore, usually fluorescein. Free analyte and labelled analyte (tracer) compete for specific antibody-binding sites in solution. The analytical signal is the value of fluorescence polarization of the fluorescein label, which corresponds to the rate of its rotation in solution. Free tracer is a small mobile molecule with a high rate of rotation. Tracer bound to antibody has a lower rate of rotation and, consequently, a higher fluorescence polarization value. Therefore the polarization value is inversely proportional to the amount of free analyte present in the sample [51].

The analytical signal in FPIA is measured like a mean characteristic of the solution (namely the fluorescence polarization value). Therefore it cannot be used for simultaneous detection of several individual analytes in a single tube. FPIA only allows detecting a signal, related to the presence of one or more compounds without discrimination. In this respect FPIA was used for urine screening for multiple drugs of abuse: cocaine metabolite, amphetamine and several barbiturates [88]; five organophosphorus pesticides in vegetable and environmental water samples [89]; multiple (fluoro)quinolone antibiotics in food samples [90]; sulphamerazine, sulphamethazine and sulphadiazine in honey and chicken muscle [91] and zearalenone and ochratoxin A in grain [92].

Capillary electrophoretic immunoassay allows combining separation of analytes from each other and from sample matrix, with high specificity of antibodies and sensitive detection such as laser-induced fluorescence detection [93,94]. Separation of the unbound fluorescent tracer and the antibody—tracer complex occurs as a result of the electrophoretic forces. Laser-induced fluorescence detection allows obtaining peak areas proportional to the fluorescein labelled analyte concentrations [51]. With cheap compact equipment, simple sample preparation, minimum handling steps and fast separation process, capillary electrophoresis with fluorescent labels become to be close to rapid test definition; laser excitation provides high sensitivity.

The advantage of capillary electrophoretic immunoassay is the integration of the separation step in this technique. Thus, a simultaneous multicomponent analysis is possible due to the high resolving power of capillary electrophoresis. Standard or sample is added to a mixture of specific antibodies and tracers. This approach was applied in the clinical assay area for detection of some drugs and drugs of abuse. Simultaneous determination of methotrexate and vancomycin was described [95]. Multianalyte capillary electrophoretic immunoassay was developed for methadone, opiates, benzoylecgonine (cocaine metabolite) and amphetamines [96,97]. For morphine and phencyclidine determination in urine samples, cyanine dyes were used as fluorescent labels: morphine was derivatized with Cy5, phencyclidine with Cy5.5 [98].

An integrated microchip electrophoresis system with online immunoreaction and laser-induced fluorescence detection has been developed for simultaneous determination of phenobarbital, phenytoin, carbamazepine and theophylline in human serum. The multiplexed immunoreactions between multiple antibody-immobilized glass beads with analytes and respective fluorescently labelled antigens were performed in a sample reservoir. After online incubation, the immunoreaction solution was injected into a one-way separation channel, and free fluorescently labelled antigens were separated and detected in the separation channel. With the help of glass beads, the immunocomplex cannot move into the separation channel, which simplifies the separation of fluorescently labelled antigens [99].

Another homogemeous assay that could be multiplexed is Förster resonance energy transfer (FRET), allowing to detect simultaneously several immunocomplex formation using either multiple donors or multiple acceptors. Possibility to use multiple energy donors was demonstrated by Zeng et al. [100] with red and green QDs. As energy acceptor, gold nanorods with plasmon absorption peaks matched with QDs' luminescence peaks. The system was used for multiple detections of Hepatitis B surface antigen and Hepatitis B antigen. Multiple energy acceptors application was shown also with an example of QDs; Tb^{3+} complex with long-lived luminescence was used as energy donor [101].

ACKNOWLEDGEMENT

The work was supported by Russian Science Foundation (project 14-13-00229).

REFERENCES

[1] L. Lafleur, D. Stevens, K. McKenzie, S. Ramachandran, P. Spicar-Mihalic, M. Singhal, A. Arjyal, J. Osborn, P. Kauffman, P. Yager, B. Lutz, Progress toward multiplexed sample-to-result detection in low resource settings using microfluidic immunoassay cards, Lab Chip 12 (2012) 1119–1127.

[2] L. Ivaska, J. Niemela, T. Heikkinen, T. Vuorinen, V. Peltola, Identification of respiratory viruses with a novel point-of-care multianalyte antigen detection test in children with acute respiratory tract infection, J. Clin. Virol. 57 (2013) 136–140.

[3] M. Hu, Y. Jan, Y. He, H.T. Lu, L.X. Weng, S.P. Song, C. Fan, L. Wang, Ultrasensitive, multiplexed detection of cancer biomarkers directly in serum by using a quantum dot-based microfluidic protein chip, ACS Nano 4 (2010) 488–494.

[4] S.W. Wang, L. Ge, Y. Zhang, X.R. Song, N.Q. Li, S.G. Ge, J. Yu, Battery-triggered microfluidic paper-based multiplex electrochemiluminescence immunodevice based on potential-resolution strategy, Lab Chip 12 (2012) 4489–4498.

[5] Y. Wang, N. Liu, B.A. Ning, M. Liu, Z.Y. Sun, Y. Peng, C. Chen, J. Li, Z. Gao, Simultaneous and rapid detection of six different mycotoxins using an immunochip, Biosens. Bioelectron. 34 (2012) 44–50.

[6] J. O'Mahony, M. Moloney, R.I. McConnell, E.O. Benchikh, P. Lowry, A. Furey, M. Danaher, Simultaneous detection of four nitrofuran metabolites in honey using a multiplexing biochip screening assay, Biosens. Bioelectron. 26 (2011) 4076–4081.

[7] C. Zong, J. Wu, C. Wang, H.X. Ju, F. Yan, Chemiluminescence imaging immunoassay of multiple tumor markers for cancer screening, Anal. Chem. 84 (2012) 2410–2415.

[8] F. Deiss, C.N. LaFratta, M. Symer, T.M. Blicharz, N. Sojic, D.R. Walt, Multiplexed sandwich immunoassays using electrochemiluminescence imaging resolved at the single bead level, J. Amer. Chem. Soc. 131 (2009) 6088–6089.

[9] M.S. Wu, H.W. Shi, L.J. He, J.J. Xu, H.Y. Chen, Microchip device with 64-site electrode array for multiplexed immunoassay of cell surface antigens based on electrochemiluminescence resonance energy transfer, Anal. Chem. 84 (2012) 4207–4213.

[10] G. Stybayeva, O. Mudanyali, S. Seo, J. Silangcruz, M. Macal, E. Ramanculov, S. Dandekar, A. Erlinger, A. Ozcan, A. Revzin, Lens free holographic imaging of antibody

microarrays for high-throughput detection of leukocyte numbers and function, Anal. Chem. 82 (2010) 3736–3744.

[11] M. Lee, S. Lee, J.H. Lee, H.W. Lim, G.H. Seong, E.K. Lee, S.I. Chang, C.H. Oh, J. Choo, Highly reproducible immunoassay of cancer markers on a gold-patterned microarray chip using surface-enhanced Raman scattering imaging, Biosens. Bioelectron. 26 (2011) 2135–2141.

[12] Y.X. Piao, D. Lee, J. Lee, T. Hyeon, J. Kim, H.S. Kim, Multiplexed immunoassay using the stabilized enzymes in mesoporous silica, Biosens. Bioelectron. 25 (2009) 906–912.

[13] F.H. Wlans, J. Dev, M.M. Powell, J.I. Heald, Evaluation of simultaneous measurement of lutropin and follitropin with the SimulTROPIN radioimmunoassay kit, Clin. Chem. 32 (1986) 887–890.

[14] L. Wang, C.Y. Yang, W.H. Tan, Dual-luminophore-doped silica nanoparticles for multiplexed signaling, NanoLetters 5 (2005) 37–43.

[15] S.H. Hu, S.C. Zhang, Z.C. Hu, Z. Xing, X.R. Zhang, Detection of multiple proteins on one spot by laser ablation inductively coupled plasma mass spectrometry and application to immuno- microarray with element-tagged antibodies, Anal. Chem. 79 (2007) 923–929.

[16] L. Chen, X.W. Zhang, G.H. Zhou, X. Xiang, X.H. Ji, Z.H. Zheng, Z. He, H. Wang, Simultaneous determination of human enterovirus 71 and coxsackievirus B3 by dual-color quantum dots and homogeneous immunoassay, Anal. Chem. 84 (2012) 3200–3207.

[17] J. Han, H. Gao, W. Wang, Z. Wang, Z. Fu, Time-resolved chemiluminescence strategy for multiplexed immunoassay of clenbuterol and ractopamine, Biosens. Bioelectron. 48 (2013) 39–42.

[18] W.B. Shim, B.B. Dzantiev, S.A. Eremin, D.H. Chung, One-step simultaneous immunochromatographic strip test for multianalysis of ochratoxin a and zearalenone, J. Microbiol. Biotechnol. 19 (2009) 83–92.

[19] Z.B. Huang, Y. Xu, L.S. Li, Y.P. Li, H. Zhang, Q.H. He, Development of an immunochromatographic strip test for the rapid simultaneous detection of deoxynivalenol and zearalenone in wheat and maize, Food Control 28 (2012) 7–12.

[20] A.Y. Kolosova, S. De Saeger, L. Sibanda, R. Verheijen, C. Van Peteghem, Development of a colloidal gold-based lateral-flow immunoassay for the rapid simultaneous detection of zearalenone and deoxynivalenol, Anal. Bioanal. Chem. 389 (2007) 2103–2107.

[21] Y.R. Guo, S.Y. Liu, W.J. Gui, G.N. Zhu, Gold immunochromatographic assay for simultaneous detection of carbofuran and triazophos in water samples, Anal. Biochem. 389 (2009) 32–39.

[22] J. Zhu, N. Zou, D. Zhu, J. Wang, Q. Jin, J. Zhao, H. Mao, Simultaneous detection of high-sensitivity cardiac troponin I and myoglobin by modified sandwich lateral flow immunoassay: proof of principle, Clin. Chem. 57 (2011) 1732–1738.

[23] T. Le, P. Yan, J. Xu, Y. Hao, A novel colloidal gold-based lateral flow immunoassay for rapid simultaneous detection of cyromazine and melamine in foods of animal origin, Food Chem. 138 (2013) 1610–1615.

[24] L. Wang, J. Cai, Y. Wang, Q. Fang, S. Wang, Q. Cheng, D. Du, Y. Lin, F. Liu, A bare-eye-based lateral flow immunoassay based on the use of gold nanoparticles for simultaneous detection of three pesticides, Microchim. Acta 181 (2014) 1565–1572.

[25] V.M.T. Lattanzio, N. Nivarlet, V. Lippolis, S. Della Gatta, A.C. Huet, P. Delahaut, B. Granier, A. Visconti, Multiplex dipstick immunoassay for semi-quantitative determination of *Fusarium* mycotoxins in cereals, Anal. Chim. Acta 718 (2012) 99–108.

[26] X. Li, K. Wen, Y. Chen, X. Wu, X. Pei, Q. Wang, A. Liu, J. Shen, Multiplex immunogold chromatographic assay for simultaneous determination of macrolide antibiotics in raw milk, Food Anal. Methods 8 (2015) 2368–2375.

[27] B.B. Dzantiev, N.A. Byzova, A.E. Urusov, A.V. Zherdev, Immunochromatographic methods in food analysis, Trends Anal. Chem. 55 (2014) 81–93.

[28] A. Abbas, A. Brimer, J.M. Slocik, L. Tian, R.R. Naik, S. Singamaneni, Multifunctional analytical platform on a paper strip: separation, preconcentration, and subattomolar detection, Anal. Chem. 85 (2013) 3977–3983.

[29] C.Z. Li, K. Vandenberg, S. Prabhulkar, X. Zhu, L. Schneper, K. Methee, C.J. Rosser, E. Almeide, Biosens. Bioelectron. 26 (2011) 4342–4348.

[30] X. Huang, Z.P. Aguilar, H. Xu, W. Lai, Y. Xiong, Membrane-based lateral flow immunochromatographic strip with nanoparticles as reporters for detection: a review, Biosens. Bioelectron. 75 (2016) 166–180.

[31] N.A. Taranova, N.A. Byzova, V.V. Zaiko, T.A. Starovoitova, Y.Y. Vengerov, A.V. Zherdev, B.B. Dzantiev, Integration of lateral flow and microarray technologies for multiplex immunoassay: application to the determination of drugs of abuse, Microchim. Acta 180 (2013) 1165–1172.

[32] C. Swanson, A. D'Andrea, Lateral flow assay with near-infrared dye for multiplex detection, Clin. Chem. 59 (2013) 641–648.

[33] I.Q. de Oliveira, R.A. Silva, M.V. Sucupira, E.D. da Silva, A.B. Reis, G. Grimaldi, D.B. Mothé Fraga, P.S. Tavares Veras, Multi-antigen print immunoassay (MAPIA)-based evaluation of novel recombinant Leishmania infantum antigens for the serodiagnosis of canine visceral leishmaniasis, Parasites Vectors 8 (2015) 45.

[34] J. Wang, H. Ahmad, C. Ma, Q. Shi, O. Vermesh, U. Vermesh, J. Heath, A self-powered, one-step chip for rapid, quantitative and multiplexed detection of proteins from pinpricks of whole blood, Lab Chip 10 (2010) 3157–3162.

[35] T.F. McGrath, L. McClintock, J.S. Dunn, G.M. Husar, M.J. Lochhead, R.W. Sarver, F.E. Klein, J.A. Rice, K. Campbell, C.T. Elliott, Development of a rapid multiplexed assay for the direct screening of antimicrobial residues in raw milk, Anal. Bioanal. Chem. 407 (2015) 4459–4472.

[36] J.P. Meneely, K. Campbell, C. Greef, M.J. Lochhead, C.T. Elliott, Development and validation of an ultrasensitive fluorescence planar waveguide biosensor for the detection of paralytic shellfish toxins in marine algae, Biosens. Bioelectron. 41 (2013) 691–697.

[37] A.E. Urusov, A.V. Zherdev, A.V. Petrakova, E.G. Sadykhov, O.V. Koroleva, B.B. Dzantiev, Rapid multiple immunoenzyme assay of mycotoxins, Toxins 7 (2015) 238–254.

[38] R. Hu, T. Liu, X.B. Zhang, Y. Yang, T. Chen, C. Wu, Y. Liu, G. Zhu, S. Huan, T. Fu, W. Tan, DLISA: a DNAzyme-based ELISA for protein enzyme-free immunoassay of multiple analytes, Anal. Chem. 87 (2015) 7746–7753.

[39] M. Magliulo, P. Simoni, M. Guardigli, E. Michelini, M. Luciani, R. Lelli, A. Roda, A Rapid Multiplexed chemiluminescent immunoassay for the detection of *Escherichia coli* O157:H7, *Yersinia enterocolitica*, *Salmonella typhimurium*, and *Listeria monocytogenes* pathogen bacteria, J. Agric. Food Chem. 55 (2007) 4933–4939.

[40] C. Lopez-Muedano, R.S. Kirton, K.D. Kumble, A.J. Taberner, Inexpensive optical system for microarray ELISA, Talanta 100 (2012) 405–409.

[41] A.G. Gehring, D.M. Albin, S.A. Reed, S.I. Tu, J.D. Brewster, An antibody microarray, in multiwell plate format, for multiplex screening of foodborne pathogenic bacteria and biomolecules, Anal. Bioanal. Chem. 391 (2008) 497–506.

[42] C. Desmet, G.C. Le Goff, J.C. Bres, D. Rigal, L.J. Blum, C.A. Marquette, Multiplexed immunoassay for the rapid detection of anti-tumor-associated antigens antibodies, Analyst 136 (2011) 2918–2924.

[43] D. Saha, D. Acharya, D. Roy, D. Shrestha, T.K. Dhar, Simultaneous enzyme immunoassay for the screening of aflatoxin B1 and ochratoxin A in chili samples, Anal. Chim. Acta 584 (2007) 343–349.

[44] E. Njumbe Ediage, J.D. Mavungu, I.Y. Goryacheva, C. Peteghem, S. De Saeger, Multiplex flow-through immunoassay formats for screening of mycotoxins in a variety of food matrices, Anal. Bioanal. Chem. 403 (2012) 265–278.

[45] N.A. Burmistrova, T.Y. Rusanova, N.A. Yurasov, I.Y. Goryacheva, S. De Saeger, Multi-detection of mycotoxins by membrane based flow-through immunoassay, Food Control 46 (2014) 462–469.

[46] E. Schneider, V. Curtui, C. Seidler, R. Dietrich, E. Usleber, E. Martlbauer, Rapid methods for deoxynivalenol and other trichothecenes, Toxicol. Lett. 153 (2004) 113–121.

[47] Q.H. He, Y. Xu, D. Wang, M. Kang, Z.B. Huang, Y.P. Li, Simultaneous multiresidue determination of mycotoxins in cereal samples by polyvinylidene fluoride membrane based dot immunoassay, Food Chem. 134 (2012) 507–512.

[48] C. Zhang, Y. Zhang, S. Wang, Development of multianalyte flow-through and lateral-flow assays using gold particles and horseradish peroxidase as tracers for the rapid determination of carbaryl and endosulfan in agricultural products, J. Agric. Food Chem. 54 (2006) 2502–2507.

[49] L. Sibanda, S. De Saeger, I. Barna-Vetro, C. Van Peteghem, Development of a solid-phase cleanup and portable rapid flowthrough enzyme immunoassay for the detection of ochratoxin A in roasted coffee, J. Agric. Food Chem. 50 (2002) 6964–6967.

[50] A. Pal, T.K. Dhar, An analytical device for on-site immunoassay. Demonstration of its applicability in semiquantitative detection of aflatoxin B1 in a batch of samples with ultrahigh sensitivity, Anal. Chem. 76 (2004) 98–104.

[51] I.Y. Goryacheva, S. De Saeger, S.A. Eremin, C. Van Peteghem, Immunochemical methods for rapid mycotoxin detection: evolution from single to multiple analyte screening: a review, Food Addit. Contam. 24 (2007) 1169–1183.

[52] I.Y. Goryacheva, S. De Saeger, B. Delmulle, M. Lobeau, S.A. Eremin, I. Barna-Vetró, C. Van Peteghem, Simultaneous non-instrumental detection of aflatoxin B1 and ochratoxin A using a clean-up tandem immunoassay column, Anal. Chim. Acta 590 (2007) 118–124.

[53] N.V. Beloglazova, I.Y. Goryacheva, T.Y. Rusanova, N.A. Yurasov, R. Galve, M. Pilar Marco, S. De Saeger, Gel-based immunotest for simultaneous detection of 2,4,6-trichlorophenol and ochratoxin A in red wine, Anal. Chim. Acta 672 (2010) 3–8.

[54] W. Jiang, N.V. Beloglazova, Z. Wang, H. Jiang, K. Wen, S. de Saeger, P. Luo, Y. Wu, J. Shen, Development of a multiplex flow-through immunoaffinity chromatography test for the on-site screening of 14 sulfonamide and 13 quinolone residues in milk, Biosens. Bioelectron. 66 (2015) 124–128.

[55] B. Weigl, G. Domingo, P. LaBarre, J. Gerlach, Towards non- and minimally instrumented, microfluidics-based diagnostic devices, Lab Chip 8 (2008) 1999–2014.

[56] P. Yager, G.J. Domingo, J. Gerdes, Point-of-care diagnostics for global health, Annu. Rev. Biomed. Eng. 10 (2008) 107–144.

[57] C.D. Chin, V. Linder, S.K. Sia, Lab-on-a-chip devices for global health: past studies and future opportunities, Lab Chip 7 (2007) 41–57.

[58] W.C. Yang, M. Yu, X.H. Sun, A.T. Woolley, Microdevices integrating affinity columns and capillary electrophoresis for multi-biomarker analysis in human serum, Lab Chip 10 (2010) 2527–2533.

[59] J.V. Jokerst, A. Raamanathan, N. Christodoulides, P.N. Floriano, A.A. Pollard, G.W. Simmons, J. Wong, C. Gage, W.B. Furmaga, S.W. Redding, J.T. McDevitt, Nano-bio-chips for high performance multiplexed protein detection: determinations of cancer biomarkers in serum and saliva using quantum dot bioconjugate labels, Biosens. Bioelectron. 24 (2009) 3622–3629.

[60] B.S. Lee, Y.U. Lee, H.S. Kim, T.H. Kim, J. Park, J.G. Lee, J. Kim, H. Kim, W.G. Lee, Y.K. Cho, Fully integrated lab-on-a-disc for simultaneous analysis of biochemistry and immunoassay from whole blood, Lab Chip 11 (2011) 70–78.

[61] C.D. Chin, T. Laksanasopin, Y.K. Cheung, D. Steinmiller, V. Linder, H. Parsa, J. Wang, H. Moore, R. Rouse, G. Umviligihozo, E. Karita, L. Mwambarangwe, S.L. Braunstein, J. van de Wijgert, R. Sahabo, J.E. Justman, W. El-Sadr, S.K. Sia, Microfluidics-based diagnostics of infectious diseases in the developing world, Nat. Med. 17 (2011) 1015–1019.

[62] B.S. Lee, J.N. Lee, J.M. Park, J.G. Lee, S. Kim, Y.K. Cho, C. Ko, A fully automated immunoassay from whole blood on a disc, Lab Chip 9 (2009) 1548–1555.

[63] J. Siegrist, R. Gorkin, M. Bastien, G. Stewart, R. Peytavi, H. Kido, M. Bergeron, M. Madou, Validation of a centrifugal microfluidic sample lysis and homogenization platform for nucleic acid extraction with clinical samples, Lab Chip 10 (2010) 363–371.

[64] H. Chen, L. Wang, P.C.H. Li, Nucleic acid microarrays created in the double-spiral format on a circular microfluidic disk, Lab Chip 8 (2008) 826–829.

[65] G. Suarez, Y.H. Jin, J. Auerswald, S. Berchtold, H.F. Knapp, J.M. Diserens, Y. Leterrier, J.A.E. Manson, G. Voirin, Lab-on-a-chip for multiplexed biosensing of residual antibiotics in milk, Lab Chip 9 (2009) 1625–1630.

[66] N.A. Taranova, A.N. Berlina, A.V. Zherdev, B.B. Dzantiev, 'Traffic light' immunochromatographic test based on multicolor quantum dots for the simultaneous detection of several antibiotics in milk, Biosens. Bioelectron. 63 (2015) 255–261.

[67] E. Song, M. Yu, Y. Wang, W. Hu, D. Cheng, M.T. Swihart, Y. Song, Multi-color quantum dot-based fluorescence immunoassay array for simultaneous visual detection of multiple antibiotic residues in milk, Biosens. Bioelectron. 72 (2015) 320–325.

[68] N.V. Beloglazova, E.S. Speranskaya, A. Wu, Z. Wang, M. Sanders, V. Goftman, D. Zhang, I.Y. Goryacheva, S. De Saeger, Novel multiplex fluorescent immunoassays based on quantum dot nanolabels for mycotoxins determination, Biosens. Bioelectron. 62 (P) (2014) 59–65.

[69] E. Panfilova, A. Shirokov, B. Khlebtsov, L. Matora, N. Khlebtsov, Multiplexed dot immunoassay using Ag nanocubes, Au/Ag alloy nanoparticles, and Au/Ag nanocages, Nano Res. 5 (2012) 124–134.

[70] C. Wang, F. Hou, Y. Ma, Simultaneous quantitative detection of multiple tumor markers with a rapid and sensitive multicolor quantum dots based immunochromatographic test strip, Biosens. Bioelectron. 68 (2015) 156–162.

[71] G.J. Worsley, S.L. Attree, J.E. Noble, A.M. Horgan, Rapid duplex immunoassay for wound biomarkers at the point-of-care, Biosens. Bioelectron. 34 (2012) 215–220.

[72] N.V. Beloglazova, O.A. Goryacheva, E.S. Speranskaya, T. Aubert, I.Y. Goryacheva, S. De Saeger, Silica-coated liposomes loaded with quantum dots as labels for multiplex fluorescent immunoassay, Talanta 134 (2015) 120–125.

[73] H. Shi, E. Sheng, L. Feng, L. Zhou, X. Hua, M. Wang, Simultaneous detection of imidacloprid and parathion by the dual-labeled time-resolved fluoroimmunoassay, Environ. Sci. Pollut. Res. 22 (2015) 14882–14890.

[74] E. Sheng, H. Shi, L. Zhou, X. Hua, L. Feng, T. Yu, M. Wang, Dual-labeled time-resolved fluoroimmunoassay for simultaneous detection of clothianidin and diniconazole in agricultural samples, Food Chem. 192 (2016) 525–530.

[75] F.B. Wu, S.Q. Han, T. Xu, Y.F. He, Sensitive time-resolved fluoroimmunoassay for simultaneous detection of serum thyroid-stimulating hormone and total thyroxin with Eu and Sm as labels, Anal. Biochem. 314 (2003) 87–96.

[76] R. Chong, J.E.R. Rho, H.J. Yoon, T.H.D. Rho, P.S. Park, Y.H. Kim, J.H. Lee, 1,1′-Oxalyldiimidazole chemiluminescent enzyme immunoassay capable of simultaneously sensing multiple markers, Biosens. Bioelectron. 32 (2012) 19–23.

[77] T. Xu, N. Liu, J. Yuan, Z. Ma, Triple tumor markers assay based on carbon–gold nanocomposite, Biosens. Bioelectron. 70 (2015) 161–166.

[78] K. Braeckmans, S.C. De Smedt, M. Leblans, R. Pauwels, J. Demeester, Encoding microcarriers: present and future technologies, Nat. Rev. Drug Discov. 1 (2002) 447–456.

[79] Y. Zhao, X. Zhao, C. Sun, J. Li, R. Zhu, Z. Gu, Encoded silica colloidal crystal beads as supports for potential multiplex immunoassay, Anal. Chem. 80 (2008) 1598–1605.

[80] R. Wilson, A.R. Cossins, D.G. Spiller, Encoded microcarriers for high-throughput multiplexed detection, Angew. Chem. Int. Ed. 45 (2006) 6104–6117.

[81] P.K. Horan, J.W. Kappler, Automated fluorescent analysis for cytotoxicity assays, J. Immunol. Meth. 18 (1977) 309–316.

[82] B. Brando, D. Barnett, G. Janossy, F. Mandy, B. Autran, G. Rothe, B. Scarpati, G. D'Avanzo, J.L. D'Hautcourt, R. Lenkei, G. Schmitz, A. Kunkl, R. Chianese, S. Papa, J.W. Gratama, Cytofluorometric methods for assessing absolute numbers of cell subsets in blood. European Working Group on Clinical Cell Analysis, Cytom. Comms. Clin. Cytom. 42 (2000) 327–346.

[83] V.V. Krishhan, I.H. Khan, P.A. Luciw, Multiplexed microbead immunoassays by flow cytometry for molecular profiling: basic concepts and proteomics applications, Crit. Rev. Biotechn. 29 (2009) 29–43.

[84] K. Xu, Y. Sun, W. Li, J. Xu, B. Cao, Y. Jiang, T. Zheng, J. Li, D. Pan, Multiplex chemiluminescent immunoassay for screening of mycotoxins using photonic crystal microsphere suspension array, Analyst 139 (2014) 771–777.

[85] H.W. Yu, I.S. Kim, R. Niessner, D. Knopp, Multiplex competitive microbead-based flow cytometric immunoassay using quantum dot fluorescent labels, Anal. Chim. Acta 750 (2012) 191–198.

[86] D.M. Bruls, T.H. Evers, J.A.H. Kahlman, P.J.W. van Lankvelt, M. Ovsyanko, E.G.M. Pelssers, J.J.H.B. Schleipen, F.K. de Theije, C.A. Verschuren, T. van der Wijk, J.B.A. van Zon, W.U. Dittmer, A.H.J. Immink, J.H. Nieuwenhuis, M.W.J. Prins, Rapid integrated biosensor for multiplexed immunoassays based on actuated magnetic nanoparticles, Lab Chip 9 (2009) 3504–3510.

[87] H. Wang, Y. Li, A. Wang, M. Slavik, Rapid, sensitive, and simultaneous detection of three foodborne pathogens using magnetic nanobead–based immunoseparation and quantum dot–based multiplex immunoassay, J. Food Protect. 74 (2011) 2039–2047.

[88] D.L. Colbert, M. Childerstone, Multiple drugs of abuse in urine detected with a single reagent and fluorescence polarization, Clin. Chem. 33 (1987) 1921–1923.

[89] Z.L. Xu, Q. Wang, H.T. Lei, S.A. Eremin, Y.D. Shen, H. Wang, R.C. Beier, J.Y. Yang, K.A. Maksimova, Y.M. Sun, A simple, rapid and high-throughput fluorescence polarization

immunoassay for simultaneous detection of organophosphorus pesticides in vegetable and environmental water samples, Anal. Chim. Acta 708 (2011) 123–129.

[90] T. Mi, Z. Wang, S.A. Eremin, J. Shen, S. Zhang, Simultaneous determination of multiple (fluoro)quinolone antibiotics in food samples by a one-step fluorescence polarization immunoassay, J. Agric. Food Chem. 61 (2013) 9347–9355.

[91] Z. Wang, S. Zhang, S. Ding, S.A. Eremin, J. Shen, Simultaneous determination of sulphamerazine, sulphamethazine and sulphadiazine in honey and chicken muscle by a new monoclonal antibody-based fluorescence polarisation immunoassay, Food Addit. Contam. A 25 (2008) 574–582.

[92] A.P. Bondarenko, S.A. Eremin, Determination of zearalenone and ochratoxin A mycotoxins in grain by fluorescence polarization immunoassay, J. Anal. Chem. 67 (2012) 790–794.

[93] J.J. Bao, Capillary electrophoretic immunoassays, J. Chromatogr. B 699 (1997) 463–480.

[94] W.S.B. Yeung, G.A. Luo, Q.G. Wang, J.P. Ou, Capillary electrophoresis-based immunoassay, J. Chromatogr. B 797 (2003) 217–228.

[95] Y. Suzuki, H. Arakawa, M. Maeda, The immunoassay of methotrexate by capillary electrophoresis with laser-induced fluorescence detection, Anal. Sci. 19 (2003) 111–115.

[96] J. Caslavska, D. Allemann, W. Thormann, Analysis of urinary drugs of abuse by a multianalyte capillary electrophoretic immunoassay, J. Chromatogr. A 838 (1999) 197–211.

[97] W. Thormann, J. Caslavska, A. Ramseier, C. Siethoff, Multianalyte capillary electrophoresis assays for screening and confirmation of urinary drugs of abuse, J. Microcolumn Sep. 12 (2000) 13–24.

[98] F.T.A. Chen, R.A. Evangelista, Feasibility studies for simultaneous immunochemical multianalyte drug assay by capillary electrophoresiswith laser-induced fluorescence, Clin. Chem. 40 (1994) 1819–1822.

[99] Y. Huang, S. Zhao, M. Shi, H. Liang, One-way multiplexed immunoassay strategy for simultaneous determination of multi-analytes by microchip electrophoresis, Analyst 136 (2011) 2119–2124.

[100] Q. Zeng, Y. Zhang, X. Liu, L. Tu, X. Kong, H. Zhang, Multiple homogeneous immunoassays based on a quantum dots–gold nanorods FRET nanoplatform, Chem. Commun. 48 (2012) 1781–1783.

[101] K.D. Wegner, Z. Jin, S. Linden, T.L. Jennings, N. Hildebrandt, Quantum-dot-based Förster resonance energy transfer immunoassay for sensitive clinical diagnostics of low-volume serum samples, ACS Nano 7 (2013) 7411–7419.

Chapter 6

Prospective Materials for Rapid Tests: Examples of Multifunctionality and Multiplexity

Chapter Outline

1. INTRODUCTION

This chapter presents two perspective directions for development of modern rapid tests. The first is development of bifunctional solid support, combining functions of substrate for immunoreagent binding, as well as development of active signal transducer and modifier. The later allows to use new principles of signal generation and enhancement. A bright example of such bifunctional platforms is photonic crystal fibres (PCFs). An exploitation of PCF as a sensitive element of biosensor allows one to get an instant optical response to the change of analysed medium's refractive index, absorption, luminescence and scattering coefficients of the medium. Such systems in perspective could be efficient solutions for clinical and food control rapid tests development due to their high sensitivity, small size and low-cost instrumentation. The diversity of unusual features of PCF leads to an increase of possibilities for new and improved sensors.

The second direction is development of enhanced labels. Selected examples of nanostructures multiloaded with quantum dots (QDs) to illustrate novel concepts and promising labels combined with the ability of QDs to produce luminescent, electrochemical or electrochemiluminescent (ECL) signals and the

Comprehensive Analytical Chemistry, Vol. 72. http://dx.doi.org/10.1016/bs.coac.2016.03.010

high carrier capacity of different carriers are presented. Approaches described include multiple QD-containing nanocomposites with organic (dendrimers and layer-by-layer (LBL) hollow microcapsules) and inorganic (carbon nanotubes and silica nanoparticles) carriers and also carriers of combined nature.

2. PHOTONIC CRYSTAL FIBRES FOR BIOSENSING

Bifunctional platforms, which play roles of solid support for immunoreagent binding, and also act as ф signal transducer, are promising tools for biosensing development. The most popular example of such setting is electrodes with immobilised immunoreagents, acting as a conductor of electrons [1,2]. To realize rapid electrochemical tests, there are several demands to be satisfied such as high sensitivity, multitarget detection and integration of electrochemical immunosensors with greater capabilities for out-of-lab application. Considerable efforts have been devoted towards the way to immobilize the recognition layer onto electrodes and the way to amplify the electrochemical signal of the binding event between the antibody and its antigen [3]. Another interesting type is application of photonic crystal waveguides (PCW) and PCFs for optical signal transfer and modulation. The use of photonic crystals for optical sensing purposes is an open research field.

A photonic crystal has a periodic dielectric structure with a periodicity of the order of a wavelength, characterized by the presence of a photonic band gap [4]. Photonic crystals can be classified on the basis of dielectric constant periodicity. In 1D crystals the refractive index periodically varies in one spatial direction. Such photonic crystals consist of parallel layers of different materials with different refractive indices and can manifest their properties in the only direction, perpendicular to the layers. Similarly, 2D- and 3D-photonic crystals are the structures where the dielectric constant varies along two or three directions, respectively. 2D-photonic crystals can be modelled by rods of dielectric in air or holes of air in dielectric material, they have the refractive index that periodically changes in two spatial directions, and they can also manifest their properties in two spatial directions. In 3D- photonic crystals the refractive index periodically changes in three spatial directions, and hence they are able to exhibit their properties in three coordinates; they can be represented as arrays of spatial bodies (spheres, cubes and so on), arranged in a 3D-crystal lattice [5].

Obtained experimental data demonstrates ability for use of PCWs as a 'smart' optical element, which gives optical response to the change of injected medium parameters. Optical waveguiding biosensors have a number of advantages over mechanical and electrical biosensors. One of the main advantages is that these biochemical sensors do not modify neither destruct the measured compounds on its surface or in the surrounding environment. They are sensitive and accurate due to the high value achievable for the signal-to-noise ratio and are immune to any electromagnetic interferences. In this kind of sensing systems, the optical path is well defined by the technological fabrication process. Optical waveguiding sensors are fast response devices, to be used in real-time measurements with multiplexed capability [4]. So-called point sensors (or physical sensors) are being used for detecting or monitoring (and even for controlling) a myriad of parameters such as temperature, strain, vibrations, pressure, refractive index, humidity, and so on.

Among variability of optical waveguides 2D-photonic crystals, namely photonic crystal fibres, are the most promising for rapid tests development geometry, because of the ease to operate, small sample volume and relatively simple read out. PCFs, also known as microstructured or holey optical fibres, are considered a major breakthrough in optical fibre technology. PCFs are characterized by a pattern of microscopic voids in the transverse plane that runs all over the waveguide [6]. Due to their holey structure PCFs have unique guiding mechanisms and optical modal properties that are not possible to achieve with conventional optical fibres [7].

Standard 'step index' optical fibres guide light by total internal reflection, which operates only if the core has a higher refractive index than the encircling cladding. There are a few types of PCFs and the dominant two are PCF with a solid core and hollow core (Fig. 1) PCFs [8]. Rays of light in the core, striking the interface with the cladding, are completely reflected. In 1991, the idea emerged that light could be trapped inside a hollow fibre core by creating a periodic wavelength-scale lattice of microscopic holes in the cladding glass – a 'photonic crystal' [9]. To understand how this might work, consider that all wavelength-scale periodic structures exhibit ranges of angle and colour ('stop bands') where incident light is strongly reflected. This is the origin of the colour in butterfly wings, peacock feathers and holograms. In photonic band gap materials, however, these stop bands broaden to block propagation in every direction, resulting in the suppression of all optical vibrations within the range of wavelengths spanned by the photonic band gap. Appropriately designed, the holey photonic crystal cladding, running along the entire length of the fibre, can prevent the escape of light from a hollow core. Thus, it becomes possible to escape the straitjacket of total internal reflection and trap light in a hollow fibre core surrounded by glass [9]. The idea of using a photonic band gap to trap light in a hollow core was intriguing for different applications, including biosensor development.

The waveguide effect in a PCW is implemented owing to internal reflection from the 'air–glass' periodic structure and creation of a photonic band gap. Fig. 2 presents the transmission spectrum of a hollow core PCF. The possibility of radiation propagation along the waveguide is determined by the radiation spectral composition. Optical radiation with frequencies that lie within the band gap cannot pass through the structural cladding of the waveguide; the radiation reflects from it and propagates along the PCF hollow core. Thus, the

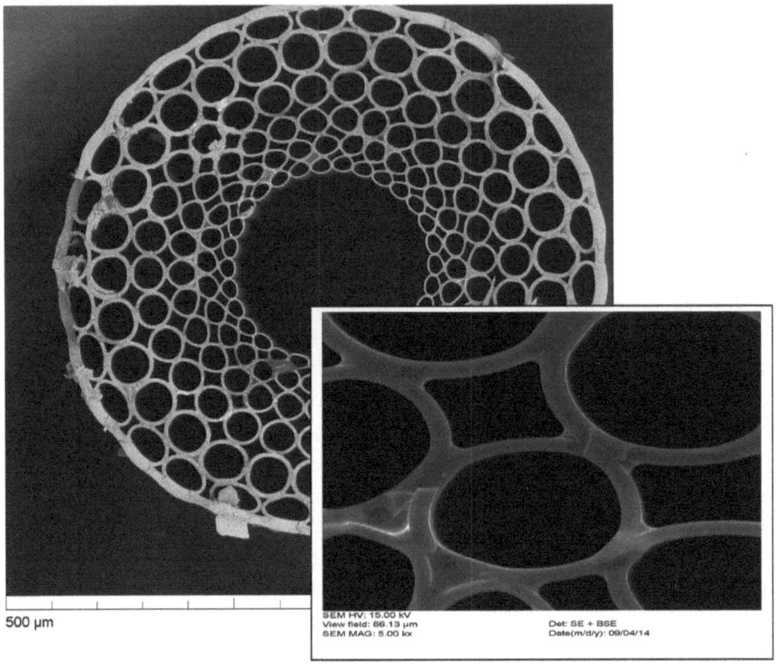

FIGURE 1 Photonic crystal fibre with hollow core.

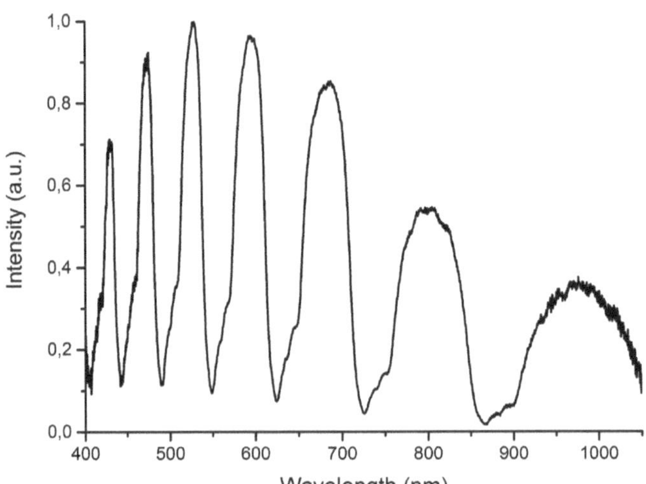

FIGURE 2 Transmission spectrum of a hollow core photonic crystal fibre.

transmission spectrum is characterized by the presence of strongly pronounced maximums and minimums corresponding to photonic band gaps of the cladding. So the PCF transmission spectrum is determined by the geometric structure of the waveguide cladding. In the case of filling of PCF hollow core with some medium, transmission spectrum becomes sensitive to the medium properties.

The composition of the material the PCF is made of is not the key feature that defines its unique properties. Geometry, mainly the design of the array of voids that form the waveguide and the use of functional materials inside the PCF voids (including postprocessing modification) play a more important role. A PCF can be made of different materials, as for example, pure silica, chalcogenide, doped and multicomponent glasses, or polymers [7].

The design of the PCF microstructure makes possible light guidance by photonic band gap effects or by the modified total internal reflection effect [6]. On the other hand, the infiltration of materials in the PCF or its postprocessing can give additional functionalities to a photonic fibre. Due to all these features, PCFs offer outstanding potential for the development of sensors. Their basic advantages include the electromagnetic noise immunity (as in many fibre-optical sensors), high sensitivity, reliability, reproducibility, wide dynamic range of measurements, the possibility of spectral and spatial multiplexing of sensitive elements, located in one or a few waveguides, short time of response to the variation in the measured quantity, small outer dimensions, the possibility to combine waveguide principles and microfluidistics [5].

PCFs are sensitive to the medium inside the PCFs, and therefore, they are good candidates to develop biochemical sensors [10]. The holey structure of PCFs offers an additional degree of freedom in the design of optical detection systems as they can be infiltrated with liquids or nanomaterials [7]. The PCF could play the role of liquid chamber. As the voids of a PCF have microscopic dimensions, the amount of liquid sample needed to carry out the sensing task is minimal, on nanolitre levels. For PCF with $\sim$300 μm hollow core diameter, 10–15 μL volume is required for filling up PCF sample of 50 mm length, while exploitation of standard photometric cuvette with 10-mm thickness requires 2–3 mL of analysed liquid for one measurement [8].

As there is a strict dependence of PCF transmission properties on spectral properties of the medium, filling up the internal channels of the fibre, that change the optical parameters of the medium, such as refractive index, scattering and absorption coefficients, induces predictable transformation of PCF transmission spectrum. A value of the refractive index inside a hollow core determines wavelengths of intensive transmission of the optical radiation through the fibre [11]. Transmission bands shift to shorter wavelengths simultaneously with increase of refractive index value. Decreasing of the transmission maxima height within entire visible range is the consequence of the presence of scattering particle. The presence of the absorbers determines the attenuation of transmission maxima within the certain wavelengths or wavelength range [5,8,12−14].

The possible classes of PCF-based sensors are presented on Fig. 3. Active PCF sensors use sensitivity of the PCF properties (transmission spectra) to the properties (refractive index) of the media inside PCF. This kind of sensors fully utilizes unique PCF optical properties. Passive sensors use PCF structure as capillary (or multicapillary) to contain sample and to guide light. Geometry properties of PCF allow to obtain several centimetres optical pathway with microlitre volume of sample. Optical properties of PCFs allow researchers to enhance optical signal (eg, luminescence). Homogeneous sensors are sensitive to parameters of media inside PCF, producing changing of properties (eg, transmission spectra) as result of presence of target compound. For heterogeneous sensors, internal surface of PCF plays (like for other heterogeneous methods) the role of modified solid support, and it is also utilized as signal transducer.

The guided light can interact directly with the solutions present in the voids of the PCF in a long section of the fibre, even metres. This gives rise to a strong light−matter interaction, hence to high sensitivity [7]. The complexity of the setups to infiltrate solutions in the PCFs is an issue that must be minimized; otherwise it will be difficult to find practical applications of PCFs for sensing. On this regard, PCF sensors can benefit from the advances in micro- and optofluidics and nanotechnology. A water-core microstructure fibre design allows nearly ideal guidance for aqueous sensing applications. The total internal reflection by a microstructured silica−air cladding provides robust confinement of light in a fluid-filled core, if the average cladding index is sufficiently below the index of water [15]. The use of PCF in optical sensing has generated much interest and promising for test application, in particular because the fibre microstructure obviates the need for complex postprocessing [16]. There are two different ways to achieve light guidance in liquid-filled hollow core PCF, depending on how the structure is initially filled: selective core filling while leaving the cladding holes air-filled and complete filling of

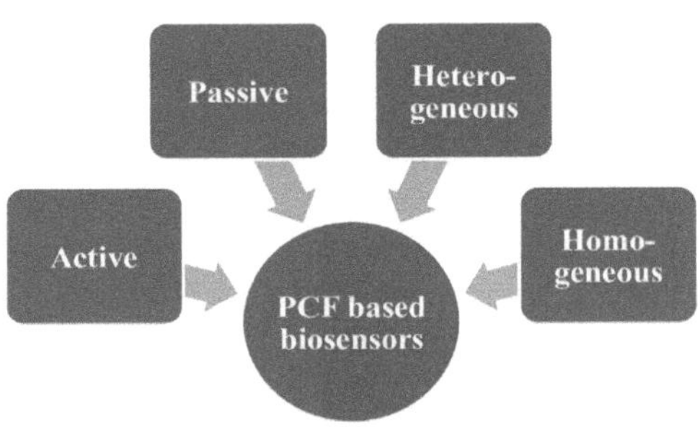

FIGURE 3 Classification of photonic crystal fibre−based biosensors.

the PCF structure, this is possible when the entire holey structure is homogeneously filled up along its length with a filling medium whose refractive index is lower than that of the PCF material.

For absorption-based sensors, PCFs with hollow core maximize light–matter interactions over much longer path-lengths than in conventional sample cells, offering an ideal environment for optical spectroscopy. Interestingly, the different types of PCF available allow the probing both of bulk solution samples and of molecules close to the internal fibre surfaces. For the first case, for example, spectra of radiation, passed through the PCF samples after filling the channels with sample could be sensitive to analyte concentration and so could show different changing, such as reduction of the peaks in transmittance spectra (short wavelength peak in the presence of riboflavin) [5] or peak shifts. For the last case surface modification through biotin–avidin interaction [17], immobilization of double-stranded DNA on the surface [18,19], or modification with antibodies [20–22] are described.

For luminescence-based sensors, in a pioneering experiment, Jensen et al. used light guidance in the glass cladding structure of a liquid-filled PCF hollow core to detect Cy5-labelled DNA molecules in submicrolitre volumes of aqueous solution [23]. Soon thereafter, Cordeiro et al. presented a PCF with a novel microstructured core design which enhanced the evanescent field overlap with infiltrated liquid samples [16,24]. Due to the high degree of achievable selectivity and sensitivity, fluorescence-based sensing has become a widely used methodology in (bio)analytical chemistry. There has been great interest in using PCF for fluorescence studies, where long interaction lengths can be achieved even for very small sample volumes. In pioneering studies, Konorov et al. [13] measured the fluorescence of thiacarbocyanine dye infiltrated into the cladding holes of a hollow core PCF. Using a solvent whose refractive index exceeded that of silica glass (dimethyl sulphoxide), an array of fluid waveguide channels was formed, in which light was guided by total internal reflection [16]. Using a suspended-core PCF made of a high refractive index lead-silicate glass, Ruan et al. enhanced the fluorescence detection sensitivity, allowing the detection of QD-labelled antibody (commercially available CdTe–ZnS QD-labelled goat F(ab′)2 antimouse IgG conjugate) down to the 1 nM level using near infrared light [25].

It was found that surface-enhanced Raman scattering (SERS) labels immobilized in the air channels of suspended-core PCF impart quantitative capacity to SERS-based PCF optofluidic sensing platform. The long optical path-length PCF integrated with SERS-active nanotags holds significant promise for sensitive quantitative chem/biomeasurements with the added benefit of small sampling volume [26]. So far, Raman spectroscopy with isolated nanoparticles in the PCF voids has been demonstrated. Such an approach allows the detection of trace amounts of molecules in liquids. The incorporation of metal dimers and trimers (clusters of two or three nanoparticles) inside the voids of PCFs may further increase the detection sensitivity, probably, down to the single-molecule level. Dimers and trimers lead to the creation of hot spots of a strong electric field in the gap between the nanoparticles which can be used to enhance the Raman scattering [27]. The feasibility of using hollow core PCF in conjunction with Raman spectroscopy has been explored for real-time monitoring of heparin concentration in serum. By nonselectively filling PCF, an enhancement of Raman signal (>90 times) was obtained from various heparin–serum mixtures filled PCFs compared to its bulk counterpart (cuvette). Promising for PCF-based tests development, the enhancement factor was directly proportional to the sample volume in the core region, which is ultimately related to the PCF length. The enhancement factor linearly increased with the PCF length in the range of 3–8 cm, and tends to saturate for higher values of PCF length [28].

3. NANOSTRUCTURES MULTILOADED WITH QDs

Luminescent QDs have shown great potential because of their unique advantages: nanoscale size similar to proteins and smaller than most of other nanoparticles with substantial emission, broad excitation spectra for multicolour imaging, robust, narrow-band luminescence and versatility in surface modification. Firstly used for bioimaging, they later became useful tools for immunoassay in the traditional microtitre plate format (fluorescent-linked immunoassay, FLISA) and after in sensors and rapid tests Chapter 3, "Formats of rapid immunotests – current-day formats, perspectives, pros and cons" by Goryacheva [89], Section 3.6.1. The interest to this kind of labels for analytical applications could be even wider because the related toxicity (the main restriction of QDs application for in vivo imaging) is not so critical for in vitro goals.

Numerous labels instead of single one involved in each molecular recognition event could amplify significantly the analytical signal output [29]. Because of relatively small size (comparing with other nanoparticles and luminescence nanostructures), physical and chemical stability and variation of surface properties, QD presents an ideal example of the label for multiplexing of different ways. The combination of association of numerous QDs into one label with a highly sensitive detection of each QD enables to detect analytes at low concentrations. One advantage of enhanced labels is the possibility to use the same signal-generation and signal-transduction settings as for the single one.

To be a suitable label for bioimaging and immunoassay, the obtained nanostructures should be water soluble. So if the support material is not water soluble, the label production process should not only include decoration of the carrier with QDs, but also a hydrophylization step. To improve the labels' properties, such as amount of loaded QDs; stability; water solubility and simplicity for bioconjugation, labels with enhanced complexity (several types of QDs, multiple biofunctionality, additional magnetic properties) is expected to open new opportunities in immunochemical applications.

3.1 Carbon Nanotubes

High attention for carbon nanostructures related with variety of properties and applications. Carbon nanoparticles – which are amorphous – are used, for example, as black-coloured labels for lateral flow immunoassay (Chapter 3: Formats of rapid immunotests – current-day formats, perspectives, pros and cons by Goryacheva [89], Section 3.5.2) [30]. Carbon nanosystems with defined structures, such as carbon nanosheets, graphenes, carbon nanotubes (CNTs) are used based on properties of their electronic structures and energetic parameters (fluorescence resonance energy transfer (FRET)-based application, for example, Chapter 2, "Rapid tests progress through the years" by Goryacheva [90], Section 2.3.3.) and geometry.

Carbon nanotubes (CNTs), because of unique mechanical, physical and chemical properties, show great potential applications in various fields, including molecular electronics, medical chemistry and biomedical engineering [31,32]. CNTs can be functionalized to achieve improved properties and functions, such as good biocompatibility and biomolecular recognition capabilities [33]. CNTs exhibit some advantages as matrix to synthesize nanohybrids. First, CNTs have a large surface area to load nanoparticles of one or different properties in a one-dimensional direction, preventing aggregation. Furthermore, they can be further decorated with biomolecules, polymers and so on. [34]. CNTs are widely used in preconcentration systems [35], as labels in electrochemical systems [36]. A general review of the use of CNTs as detection labels has been published by Scida et al. [37].

Recently, many works have reported the synthesis of CNT–QD nanohybrids through covalent or noncovalent approaches (Fig. 4) and their potential as candidates for multifunctional architectures, involving in vivo or in vitro fluorescent nanoprobes [38]. Both single- [39] and multiwalled [38] full-length and shortened by acidic oxidation [40] CNTs were used for nanostructures creation by variety of possibilities. Oxidation of carbon nanotubes also results in formation of carboxyl groups on the nanotube sidewalls as well as the ends [40]. The possibilities of following functionalization of obtained COOH groups are discussed in detail by [39]. Also, the COOH group itself makes the CNTs water soluble. Another way to make CNT water soluble is their modification with polar compounds, for example, dendrimers [41].

It has been widely reported that the QDs emission could be strongly suppressed when they are attached to carbon nanostructures, which may be attributed to the FRET from QD to CNT [32]. For electrochemical application CNTs have an opposite effect: Ding et al. [42] and Jie et al. [43] reported that carbon nanotubes could enhance QDs ECL by reducing the injection barrier of electrons to QDs. Moreover, a single CNT can encapsulate dozens of QDs (as the signal source) to produce amplification of an ECL signal [44]. On the other hand, the more porous structure of the composite material could facilitate

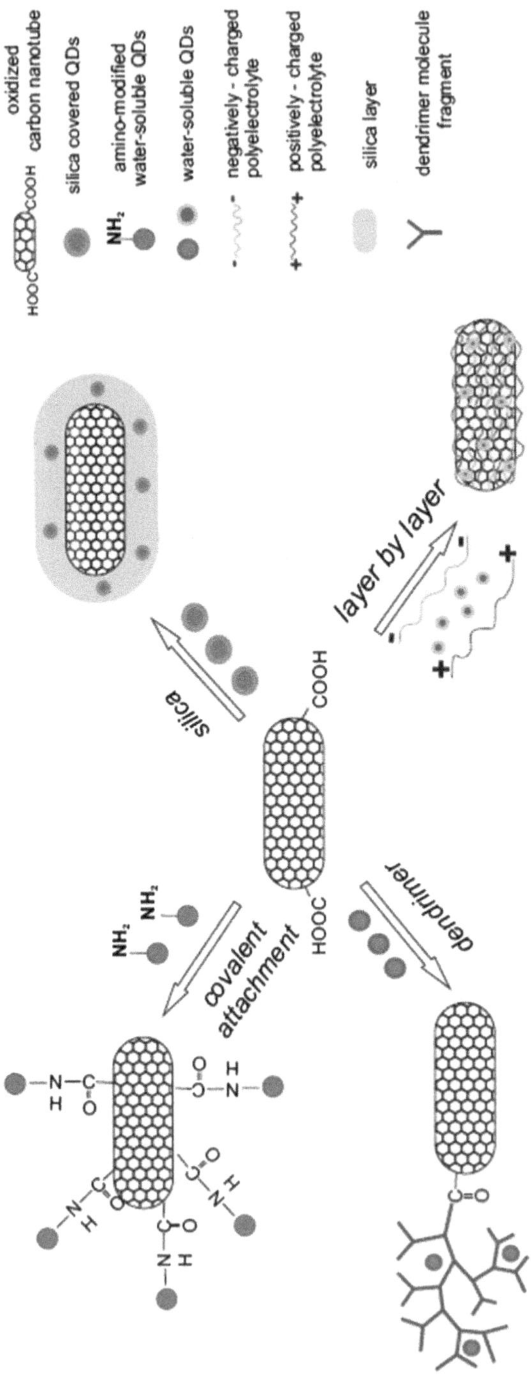

FIGURE 4 Possibilities for carbon nanotubes multiloaded with quantum dots.

the diffusion of reagents and result in ECL occurring not only at the surface but also in the bulk of the nanomaterial [42].

The amount, density and single/multiple colour of QDs functionalized onto the carbon nanotube's surface can be varied by the synthesis conditions. Functionalization of nanotubes with luminescent QDs can be accomplished through different principles and procedures by both chemical bond formation and electrostatic interaction [45–47]. An application of QDs as electrochemical label does not need ideal structures of QDs, as, for example, luminescent labels require. This gives possibility to create electrochemically active labels multiloaded with QDs via simple one-pot methods. For example, for synthesis of ZnO QDs dotted carbon nanotubes carboxyl functionalized CNTs were simply mixed with zinc acetate complex with ethylene glycol. Obtained structures were used as sensitive ECL label [48].

The more QDs attached to the CNT surface, the higher analytical signal can be expected theoretically. To increase the amount of QDs on the CNT surface, different approaches, such as dendrimer or LBL formation, were used. Modification with dendrimers can be either covalent or noncovalent; and it generally employs polyamidoamine (PAMAM) dendrimers [49].

CNTs could be the base for multifunctional nanohybrids with different properties synthesized using the LBL technique. Nanohybrid CNT with superparamagnetic iron oxide nanoparticles and CdTe QDs were synthesized via an electrostatic LBL assembly in combination with covalent connection strategy. CNT were covered with poly(allylamine hydrochloride), then superparamagnetic iron oxide nanoparticles and QDs (both with activated carboxylic groups) were bound by covalent coupling to the polymer's amino groups on the CNTs' surface [33].

Simultaneous attaching of two different-coloured QDs on nanotubes could simplify detection in the case of complex matrices by multicolour fluorescence imaging. It can be anticipated that the QDs with different luminescence colour decorated on one single CNT can be conjugated with various biological molecules and drugs (DNA, antibodies, siRNA, etc.). Therefore, the obtained multicolour-enhanced nanoprobes could be extended to more sophisticated bioapplications such as multicolour coding, multiplexed assays and multitargeted biomedical treatment by using carbon nanotubes as an internalization vehicle for carrying not only luminescent probes but also functional biomaterials [47]. Multicolour nanoprobes were prepared also by electrostatically assembling differently sized CdTe QDs on polyethylenimine-functionalized multiwalled CNT.

Structures described earlier still have limited analytical application. For detection of quenching of QDs' luminescence caused by CNTs, FRET was used. CNTs and QDs, their surfaces functionalized with oligonucleotide DNA or antibody, can be assembled into nanohybrid structures upon the addition of a target complementary oligonucleotide (DNA hybridization in the presence of target complementary oligonucleotides) or antigen (sandwich type immunocomplex formation in the presence of target analyte). The method is achieved with a detection limit of 0.2 pM DNA molecules and 0.01 nM antigen molecules. To simultaneously detect three target DNA molecules, three kinds of QDs with different emission wavelengths at 510, 555 and 600 nm were used [32].

QD luminescence suppressed by FRET through strong interaction between QD-single-stranded DNA and CNT recovered in the presence of target DNA due to decreasing the interactions between the QDs and CNTs. The recovered emission of QDs was linearly proportional to the concentration of the target in the range of 0.01–20 μM with a detection limit of 9.39 nM. Moreover, even a single-base mismatched target with the same concentration of target DNA can only recover a limited low luminescence of QDs, illustrating the good anti-interference performance of this QD/DNA/CNT system [50].

Completing the description of the CNT-QD nanocomposites as producing signal labels it is important to note the various types of CNT–QD interactions, such as charge, energy or electrons transfer which may affect the final signal in both directions: the rise and the reduction. This opens good perspectives for the further improvement of the analytical signal, but also requires a detailed study of the processes and the influence of assay conditions for each case.

3.2 LBL Microcapsules

The LBL technique is based on the sequential absorption of substances with different properties. The Layer-by-layer (LBL) technique is simple and cheap, which is extensively used in a wide variety of materials, such as polyions, metals, ceramics, nanoparticles and biological molecules. Most often these substances are synthetic or natural polycations and polyanions [51], but other compatible systems, such as streptavidin–biotin [52] could be used. The LBL technique allows to easily regulate the size and properties of obtained structures by an amount of layers, external charge and density of included labels.

In the common case electrostatic attraction between consecutively deposited, oppositely charged molecules leads to the formation of a stable film assembly. Strong electrostatic interaction between polycations and polyanions, which is the driving force for the deposition of subsequent shell layers cause an overcharge of the external layer. Firstly this strategy was used for formation of the multilayer films on flat macroscopic surfaces. Later, this technique was applied to obtain multilayers around charged colloidal particles [53]. The LBL technique of multilayer formation can use latex colloids, calcium carbonate and silica as capsule cores. The colloidal template can be removed in the dissolution step ($CaCO_3$ in acid or EDTA solution) in order to leave an empty LBL shell. The alternative approach is the direct formation of the polyelectrolyte shell without template on the emulsion droplets. If such droplets contain hydrophobic substances, this is the way for introducing them into microcapsules. For encapsulation of hydrophilic compounds, their complexes with the oppositely charged polyelectrolyte can be used as a key for their further encapsulation by the LBL technique [54].

The main general features of the LBL technique for preparation of QD-loaded microcapsules are [54]: (1) it is a facile self-assembly method avoiding complicated chemical reaction to form QD composite associates; (2) the QDs' loading, composition, density and distribution in the LBL structure can easily be controlled by the number of sequential absorption cycles and the kind of layer-formative material (eg, the nature of polymer); (3) QDs can be incorporated both in template and LBL-formed structures (Fig. 5 from [29]).

The size of the LBL assembly is determined by the size of the template and is in the range of 0.2–3 μm [55,56]. If preparing without template, size of LBL associate mostly depends on the amount of layers and the nature of polyelectrolyte. Size, topology of microspheres and also external coating can be selected to meet the requirements for their further application. Antibody attached to the outer wall of capsules can specifically bind with the targetting analyte or attach to the targeting place [54].

The number of deposited QDs was evaluated on the basis of the concentration change of the QD suspension before and after the coating step. For CdS QDs depositing on nanosized polystyrene particles based on streptavidin–biotin association, the amount of QDs on each bead was estimated 6.7×10^4 after attachment of the first layer of biotin-coated QDs and 9.2×10^4 after the fourth streptavidin layer [56].

To obtain a high QD loading, $CaCO_3$ nanoparticles were used as cores for LBL association of QDs by the electrostatic adsorption technique. Then templates were dissolved in ethylenediaminetetraacetic acid. QDs were present both in the shell and the hollow core of the LBL structure. Obtained label was further used for low-abundance protein detection. The colour of the synthesized structures was more intense after removal of the $CaCO_3$ template possibly because most of the QD could adequately show their native colour [57].

It was found that certain parameters of the LBL structure could influence QD emission and the resulting luminescence signal. Encapsulation of QD into poly-L-lysine/poly-D-glutamic polyelectrolyte capsules did not affect their spectra as long as capsule shells were terminated with a poly-L-lysine layer. A slight decrease of the fluorescence intensity was observed in the opposite case due to the presence of electron withdrawing carboxylic groups on the surface. Comparison of the QDs' fluorescence properties incorporated into LBL structures formed by synthetic and natural polyelectrolytes did not show any significant difference. In the same time it was found that encapsulation of QD within internal layers of biocompatible electrolytes reduced their cytotoxicity. On the contrary, the capsules prepared with synthetic polyelectrolytes did not show such an effect [54]. Generalova et al. [55] showed that addition of bovine serum albumin (BSA) as an external coating for the LBL structure resulted in an increased fluorescence intensity of polymer particles with QD. These results likely can be explained in terms of complex formation between BSA and

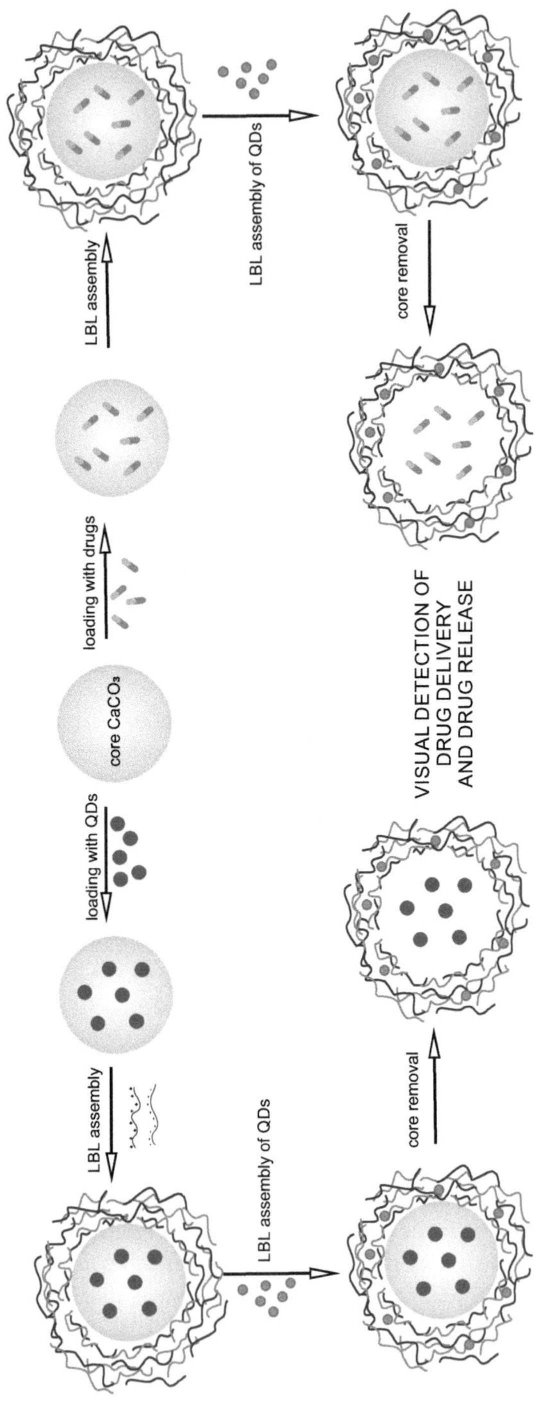

FIGURE 5 Approaches for quantum dots loading into layer-by-layer structures. *From I.Y. Goryacheva, E.S. Speranskaya, V.V. Goftman, D. Tang, S. De Saeger, Synthesis and bioanalytical applications of nanostructures multiloaded with quantum dots, Trends Anal. Chem. 66 (2015) 53–62.*

poly(sodium styrene sulphonate) resulting in more effective passivation of surface of QD, situated in the poly(sodium styrene sulphonate) layer.

Analytical signals such as photoluminescence, or ECL, or anodic-stripping voltammetry peak current are depending on the additive effect of all QDs in the label. Theoretically, the amount of QDs contributing to the analytical signal, increases with an increasing number of QD layers. Both optical density in the range of exciton peak and voltammetric response are gradually increasing as the number of QD layers increases, indicating the attachment of more and more QDs [56]. Similarly, the deposition of QDs in two layers appeared to increase the fluorescence intensity more than three times as compared to the fluorescence of a structure with QDs in one layer [55]. However too many layers might cause instability of the LBL associate due to the electrostatic adsorption capacity [57]. It was shown that the anodic-stripping voltammetry peak current significantly increased with an increasing number of QD-containing layers up to five layers, and then slightly decreased for the following additional layers. Authors [57] speculated that during LBL-structure preparation (centrifugation and washing process) QD may partially be detached from the LBL structure. In order to show a dramatic signal amplification, the detection with a single CdS QD and with an LBL associate with four layers was compared. A 17-fold electrochemical signal enhancement with the LBL label along with the detection of a 100 times smaller target analyte concentration (10 pM and 100 fM, respectively) was observed [56].

On the contrary: if the analytical signal generation is the result of interaction of a single QD, resulted in changes of its properties, no dramatic effect would occur. For example, sensitivity of Cu^{2+} detection on the base of the fluorescence quenching of QDs covered with BSA was comparable for the single QD (10 nM) and the QD-loaded LBL associate (15 nM) [55].

Although in biomedical research, the application of LBL-designed structures with QDs is based on enhanced luminescence; impressive analytical applications are mostly electrochemical in nature while a variety of principles of signal generation/amplification is used. Cai et al. [58] described an electrochemical sensor based on porous calcium carbonate microspheres covered with LBL structures containing QDs. The haem centre of haemoglobin loaded in the pores behaves as a peroxidase, reducing peroxide in the analysed solution to water. CdTe QDs helped to realize a fast and effective direct electron transfer between the haem centre of the haemoglobin molecules and the macroscopic electrode. Xiang et al. [56] and Zhang et al. [57] used QD-loaded LBL structures as labels for electrochemical immunoassay. QD-loaded LBL were used for detection of model protein IgG1 on the base of anodic-stripping voltammetric detection of Cd^{2+}, which was released under acidic conditions from the CdS QDs [51]. QDs as electrogenerated chemiluminescent labels were applied for thrombin detection at subpicomolar level using CdTe QD attached on the surface of polystyrene microbeads by the LBL technique based on the strong streptavidin—biotin association [52].

The LBL technique provides good perspectives for analytical applications because of possible variations in size and charge permeability, stability of obtained structures, distance between QDs and external surface.

3.3 Dendrimers

Dendrimers are regular tree-like highly branched macromolecules, which are receiving considerable attention for applications in chemical and biological areas owing to their numerous terminal groups that can be functionalized and conjugated with nanoparticles and molecules of interest. They are highly ordered, regularly branched and present essentially as monodisperse single compounds which develop a more globular conformation as its generation increases [59,60]. The most popular material for dendrimer is PAMAM. PAMAM dendrimers are water-soluble polycations, and they can be protonated in aqueous solution at termini (primary amines) and at branch points (tertiary amines) with the extent of protonation depending on the solution pH [61]. The amino groups on the nonmodified PAMAM can be easily bioconjugated or modified. PAMAM dendrimers of different generation are commercially available, both nonmodified (only amino groups) and modified with hydroxy-, carboxy- and trimethoxysilyl-groups.

For conjugation with QDs, dendrimers of third, fourth or fifth generation (G3, G4 and G5, respectively) containing 32, 64 and 128 primary amino-groups, correspondingly, are usually used. According to Divsar and Ju [62], CdS QDs nanocomposites prepared using two generations of dendrimers displayed similar fluorescence properties. Modification of QDs with PAMAM allows to change charge: the zeta potential of PAMAM dendrimer-conjugated QDs was slightly positive (+2.76 mV), although the zeta potential of QDs was negative (−19.3 mV) [60].

Decoration of dendrimers with QDs can be controlled by different methods, such as: photoluminescence, absorbance [60,62], high-resolution electron microscopy, Fourier transform infrared (FTIR) and zeta (ζ) potential measurement [59]. Pure dendrimers were positively charged with a zeta potential of +8 mV due to the amino groups. After loading of negatively charged CdSe/ZnS QDs on dendrimers, the zeta-potential shifted to −4 mV [63].

The ratio of QD/dendrimer is of importance. Some applications only use covering of a single QD with dendrimers to improve signal stability, to allow cell penetration, to optimize charge or to introduce groups for conjugation. Different ratios of PAMAM dendrimer to QD are reported. Akin et al. [64]. used 1 to 28 PAMAM G5 to each QD. This is a way to obtain high water solubility and positive charge of the QD surface. According to Divsar and Ju [62], each QD was stabilized by about three G4 or one G5 dendrimer. Other applications need more complex nanocomposites, consisting of multiple QDs. For this purpose, dendrimer nanoclusters were prepared by cross-linking PAMAM dendrimers with homobifunctional amine-reactive agents, such as

N-hydroxysuccinimide functionalized polyethylene glycol (NHS-PEG-NHS) (Fig. 6 from [29]). The presence of a PEG spacer arm helped to maintain the high water solubility of the formed dendrimer clusters. The nanocluster size can be regulated by changing the molar ratio between the PAMAM dendrimer and the cross-linker [65]. A molar ratio of 50:1 [NH_2]/[NHS] was found appropriate to obtain nanoclusters with an average diameter of about 150 nm. After cross-linking G5 dendrimers to larger nanoclusters, their labelling with large numbers of QDs could be easily realized [40]. Another example of creating label with multiple QDs is the individual dendrimers' covalent conjugation to the MNPs with carboxyl groups, which were then cross-linked to larger dendrimer composite by NHS−PEG−NHS. Finally, the CdSe−CdS QDs were assembled upon the dendrimer composites by covalent conjugation Fig. 7 from [66].

While it is easy to synthesize QDs in water phase and attach them to dendrimers, the synthesis of highly fluorescent QDs is generally performed in organic solvents. As for biosensing applications, QDs have to be conjugated to biological molecules. This requires the transfer of QDs into aqueous solutions which is quite challenging. PAMAM dendrimers with amino moieties are found to solubilize QDs in aqueous solvents through ligand-exchange reactions. An efficient phase-transfer reaction of QDs is possible, because of the electrostatic stabilizing feature of multiple positive groups of PAMAM dendrimers [64].

As other 'hosts', dendrimers can be a carrier for multifunctional hybrid structures. For example, by incorporating magnetic nanoparticles, superstructures with highly intense ECL, fluorescence and magnetic properties can be obtained. Furthermore, nanocomposites with excellent magnetic properties can be easily labelled, separated and immobilized [66]. Also dendrimers can tremendously increase the capacity for loading QDs on 'inert' carriers, such as CNTs, as there are a lot of terminal amino groups on the surface of the dendrimer based on which high density composites can be synthesized. Furthermore, the density of QDs on the synthesized composites can be easily controlled by varying the ratio of nanoparticle/CNT [67].

If chemical process is one of the steps, the signal generation uniformity of QDs' localization on the dendrimer surface is important for fast and easy availability for reagents. This also favours the electron transfer in the ECL reactions. For application with electron transfer processes, the amine functional groups in the dendrimers facilitate the generation of radicals and accelerate the electron-transfer process during the ECL reaction [63]. For applications as luminescent label, the significant point of dendrimers is that no luminescence quenching is reported.

For ECL applications QDs−PAMAM dendrimers can be obtained by one-pot synthesis in methanol through formation of ion pairs PAMAM-Me^{2+} and then S^{2-} addition. The CdS QDs−dendrimer nanocomposite synthesized with dendrimers of different generations (PAMAM G4 and G5) showed similar

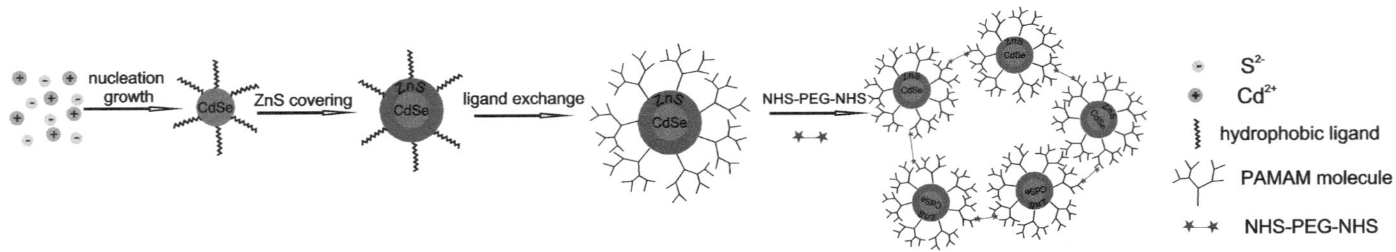

FIGURE 6 Preparation scheme of multilabel based on dendrimer-covered quantum dots. *From I.Y. Goryacheva, E.S. Speranskaya, V.V. Goftman, D. Tang, S. De Saeger, Synthesis and bioanalytical applications of nanostructures multiloaded with quantum dots, Trends Anal. Chem. 66 (2015) 53–62.*

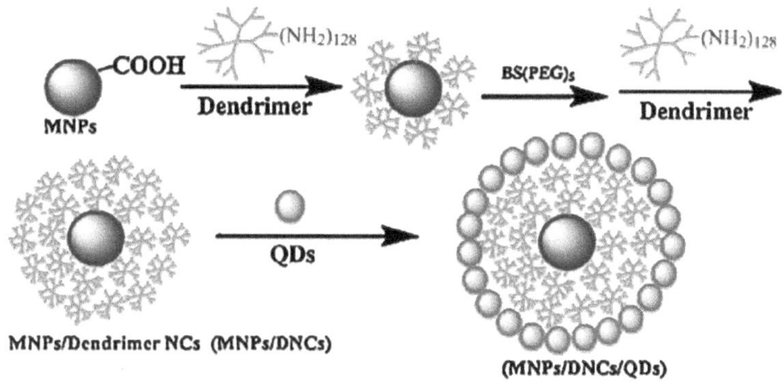

FIGURE 7 Schematic representation for the preparation of the magnetic nanoparticles (MNPs)/ PAMAM dendrimer of fifth-generation ($(NH_2)_{128}$)/CdSe/CdS QDs nanostructure. *From G. Jie, J. Yuan, J. Zhang, Quantum dots-based multifunctional dendritic superstructure for amplified electrochemiluminescence detection of ATP, Biosens. Bioelectron. 31 (2012) 69–76.*

fluorescence properties (λ(fl) = 495 nm); their maximum absorption peaks occurred at 326 and 279 nm, respectively, exhibiting average particle sizes of QDs to be 1.66 and 1.30 nm [62]. Tang et al. [68] used this approach to obtain composites of dendrimer with CdS, ZnS and PbS Fig. 8 form [68]. These dendrimer–QD composites were used for a novel multiplexed stripping voltammetric immunoassay for simultaneous detection of three biomarkers (CA 125, CA 15-3, and CA 19-9). Yamamoto et al. [69] studied factors affecting CdS QDs formations in *n*th-generation dendrimers and found that setting the Cd^{2+} ion/dendrimer ratio to be less than 2^n enabled most of the nucleation to occur inside the dendrimers, which led to stable dendrimer-encapsulated CdS QDs. It was demonstrated that higher-generation dendrimers can confine smaller QDs because they have denser surfaces.

For analytical purposes, nanocomposites PAMAM-CdSe QDs were used for detection of nitroaromatic compounds as quenchers of QDs' fluorescence [70]. A dendrimer/CdSe/ZnS QD nanocluster was fabricated and used as an ECL probe for assays of cancer cells. In comparison with the single QDs, a 13-fold enhancement in ECL signal was observed using the dendrimer–QDs nanostructure [64]. ECL of the nanocomposites dendrimer/QD/magnetic nanoparticle was efficiently quenched by gold nanoparticles, based on which a novel strategy for signal-on ECL detection of adenosine-5′-triphosphate ATP was developed [69]. An LBL assembly of QD/dendrimer multilayer systems was used as energy donor for sensitive detection of DNA hybridization by FRET with a detection limit as low as 100 fM [71].

Stripping voltammetric analysis is an analytical technique for QD-based electrochemical immunoassay that involves (1) preconcentration of a metal onto a solid electrode surface or into Hg (liquid) at negative potentials and (2) selective oxidation of each metal species during an anodic potential sweep.

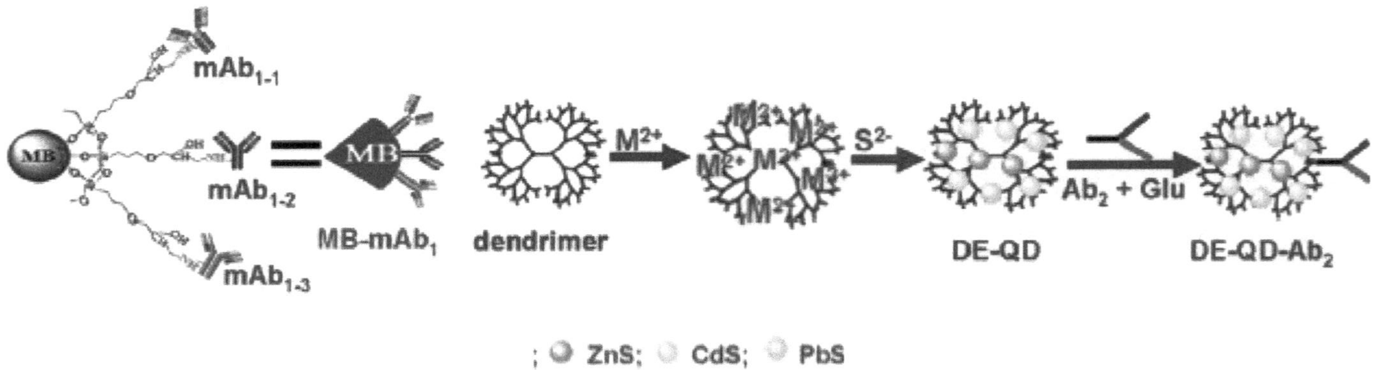

FIGURE 8 Schematic illustration of preparation process of metal-sulphide QD-loaded dendrimer for multiplexed stripping voltammetric immunoassay of biomarkers. M^{2+}: Zn^{2+}, Cd^{2+}, Pb^{2+}; $mAb_{1\text{-}1,1\text{-}2,1\text{-}3}$:primary antibiomarker antibody; Glu: glutaraldehyde; Ab_2: detection antibodies. *From D. Tang, L. Hou, R. Niessner, M. Xu, Z. Gao, D. Knopp, Multiplexed electrochemical immunoassay of biomarkers using metal sulfide quantum dot nanolabels and trifunctionalized magnetic beads, Biosens. Bioelectron. 46 (2013) 37–43.*

Metal ions, such as cadmium, copper, zinc and lead, can be stripped from the corresponding QDs under harsh conditions. The stripped metal ions can exhibit specific voltammetric characteristics at different applied potentials. Compared with conventional multienzyme labels, the metal sulphide QD-based tags are controlled to interact with the primary antibodies-functionalized magnetic beads and produce the corresponding current responses by the anodic stripping voltammetry. More importantly, the QD-encapsulated PAMAM not only contains a large number of QDs, but also favours the dissolution of QD during the stripping voltammetric measurement [68].

Dendrimers are uniformed 'building bricks' for creating nanostructures; their use is not complicated with quenching processes; they contain amino groups that can interact directly with the QD surface. But till now the dendrimer application for multiple QDs labels creation is less developed comparing to other multiloaded with QDs nanostructures.

3.4 Silica Nanoparticles

Encapsulation of hydrophobic QDs into a silica shell is a well-known approach to make QDs water soluble. The first studies were mainly devoted to cover a single QD with a SiO_2 shell [72], while the later ones focussed on SiO_2 beads with multiple fluorescent QDs inside in order to achieve a noticeable signal enhancement in ultrasensitive bioanalysis or to obtain multifunctional labels [73].

Silica coating has the following advantages compared to alternative coating strategies [73,74]: silica is inert in both aqueous and nonaqueous solvents, so the silica shell can dramatically reduce the release of harmful ions [75] and can decrease environmental interferences on the QDs' brightness. In addition, silica coating is optically transparent, and different functional groups (amines, carboxyls or methacrylate) can be easily introduced to the silica shell [76]. Moreover, silica coating reduces nonspecific adsorption on the QDs.

To encapsulate several QDs into silica beads two main methods are used for the preparation of functionalized silica nanoparticles: the Stöber and reverse microemulsion process. In both 'bottom-up' approaches, the particles are formed by self-assembly. Up to now both hydrophobic and hydrophilic QDs can be encapsulated by these approaches. The first method is a *sol–gel process* (so-called Stöber method) where QDs act as seeds for silica growth in an ethanol/water mixture [73,77]. The typical procedure of encapsulation of multiple hydrophobic QDs into SiO_2 beads includes three stages [74,78,79]: silanization of the QDs by exchange of initial hydrophobic ligands with partly hydrolysed tetraethyl orthosilicate (TEOS); condensation of the TEOS molecules attached to the QDs' surface with partially hydrolysed 3-mercaptopropyltrimethoxysilane (MPS) and formation of hydrophilic seeds; growth of a SiO_2 shell around several seeds by Stöber synthesis (Fig. 9 from [29]).

For encapsulation of hydrophilic QDs by this approach, at first the initial hydrophilic ligands are partly or completely exchanged to MPS molecules with subsequent assembly of the silanized QDs into seeds. Then the seeds are encapsulated into SiO_2 beads by Stober synthesis. In both cases the particles before the Stöber process were precoated with a silanizing agent (such as MPS or 3-aminopropyltrimethoxysilane (APTES)) to enhance the compatibility of nanoparticles with silica for growth of a uniform shell in the last stage [80].

It was shown that the quantity of QDs per bead, fluorescence intensity as well as SiO_2 shell thickness could be sufficiently varied by changing the following process parameters: QDs' and other reagents' amount and TEOS/QDs ratio [73,79]. It was found that QDs comprising SiO_2 nanocomposites with very different morphology (fibre-like, sheet-like, pearl-like, with solid or hollow sphere) could be obtained when varying the conditions during the ligand-exchange and seed-formation processes. It was proved that the photoluminescence efficiency of QDs in SiO_2 beads and so the brightness of the label itself was extremely sensitive to the incorporation conditions: in the case of both hydrophobic and hydrophilic QDs, only slow ligand exchange in the beginning provided well-ordered arrangement of TEOS or MPS on a QD surface and thus proper passivation of the surface and retention of initial quantum yield [73].

Despite the good results obtained with the Stöber method, it has some serious disadvantages such as the high requirements on purity of the reagents,

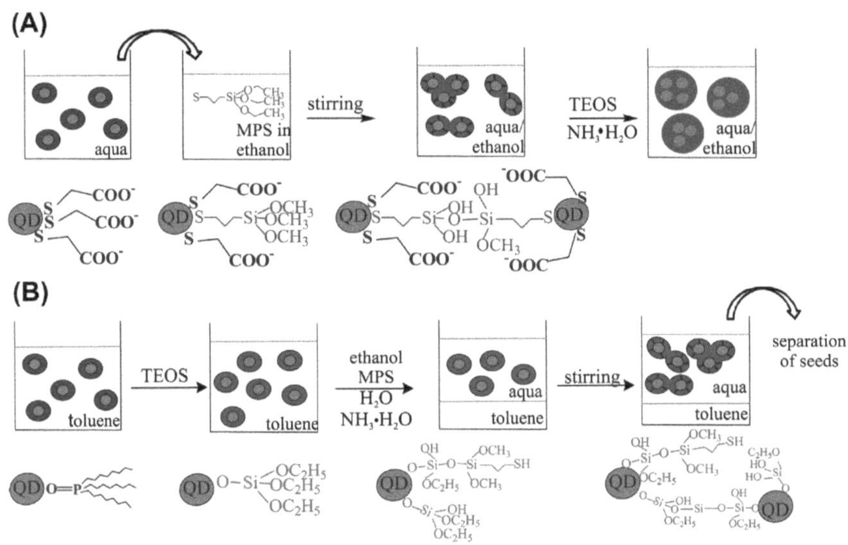

FIGURE 9 Quantum dots encapsulation into silica nanoparticles using Stöber synthesis: (A) traditional method and (B) in the presence of trioctylphosphine oxide (TEOS, tetraethyl orthosilicate; MPS, mercaptopropyltrimethoxysilane). *From I.Y. Goryacheva, E.S. Speranskaya, V.V. Goftman, D. Tang, S. De Saeger, Synthesis and bioanalytical applications of nanostructures multiloaded with quantum dots, Trends Anal. Chem. 66 (2015) 53–62.*

the difficulty and multiplicity of the preparation steps [81]. Another major limitation of the conventional Stöber process is that it is inherently incompatible with proteins when used in the bioassay. The most likely reason is the fact that the highconcentration alcohol used and the basic pH of the reaction tend to lead to partial or even full denaturation of the protein. In comparison with the Stöber approach, the second approach based on the *reverse microemulsion method* (RMM, also known as 'water-in-oil' microemulsion systems) is considered more advantageous because it is less complicated, very 'robust' against many reaction conditions and because the particle size and size dispersion can be better controlled [82]. In RMM, the micelles are used to confine the seed particles and to control the deposition of silica within the micelle [80,83]. In a typical synthesis, the nonionic surfactant (eg, Synperonic NP-5) and organic solution of QDs were dispersed in the continuous phase (eg, cyclohexane) followed by addition of TEOS and ammonia (Fig. 10 from [29]). It was proved that the original hydrophobic ligands of QDs were replaced by TEOS molecules and the QDs transferred into the hydrophilic interior of the micelles [81]. TEOS molecules underwent hydrolysis and condensation in the water phase and the QDs acted as nucleation centres. As a result, single or multi-QDs encapsulated into SiO_2 were obtained.

It was found that the size, monodispersity, multiplicity of QDs per silica particle and overall quality of the resulting QDs silica particles were governed by reaction conditions such as time, temperature, concentration of reagents and speed of stirring [81,83]. Particularly, with increasing QDs' concentration within a certain range the number of QDs per silica bead increases [83], but at too high QDs' concentration, the micellar system becomes unstable, resulting in an aggregation of the QD/silica nuclei [84]. The use of QDs with a thicker shell of wider band gap semiconductor allowed to sufficiently retain the initial brightness.

For further bioapplications, the surface of the biolabel should be decorated by functional groups such as carboxy-, thiol- or amino-. The SiO_2 surface can be modified through a sol−gel process by the use of different silanizing agents capable of hydrolysis and subsequent condensation with the existing SiO_2 network: carboxyethylsilanetriol sodium to introduce COOH groups, MPS for thiol groups, APTES and [3-(2-aminoethylamino) propyl] trimethoxysilane for NH_2. To decrease nonspecific sorption of SiO_2 beads, the same approach can be used to introduce PEG-chains into outer shelling using 2-[methoxy(polyethyleneoxy)propyl]-trimethoxysilane [73].

In the mid-2000s labels with both supermagnetic and fluorescent functionalities (bifunctional nanocomposites − BNPs) attracted intense attention due to their great potential [60,83,85]. As a rule, for preparation of such labels, iron oxides (Fe_3O_4 or Fe_2O_3) are used as magnetic nanoparticles (MNPs), QDs (or organic dyes) − as fluorescent particles. Highly luminescent QDs serve as luminescent markers, while (MNPs) can be easily manipulated under the

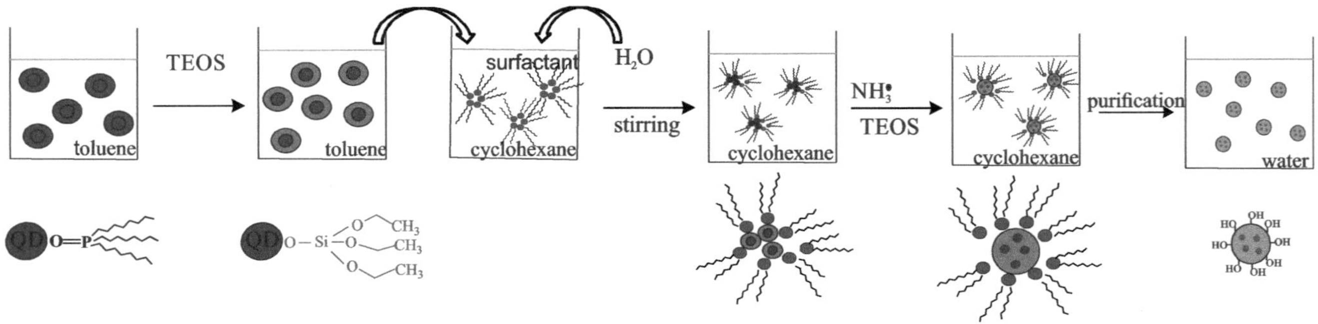

FIGURE 10 Quantum dots encapsulation into silica nanoparticles reverse microemulsion method. *From I.Y. Goryacheva, E.S. Speranskaya, V.V. Goftman, D. Tang, S. De Saeger, Synthesis and bioanalytical applications of nanostructures multiloaded with quantum dots, Trends Anal. Chem. 66 (2015) 53–62.*

external magnetic field [85]. One of the approaches for BNPs' synthesis is coating of nanoparticles with silica [83].

In the first studies, BNPs were prepared by simultaneous encapsulation of fluorescent QDs and MNPs inside SiO_2 beads by RMM [86]. It was shown that such BNPs had decreased fluorescence brightness compared to initial QDs [87]. Then the subsequent experiments elucidated that the photoluminescent properties of BNPs were very sensitive to the distance between the magnetic and fluorescent nanocrystals in BNPs and that BNPs' fluorescence brightness could be sufficiently improved by suppression of interaction between the two particle types [85]. QDs and MNPs can be spatially separated by a SiO_2 shell [88]. Different types of BNPs embedded into SiO_2 beads and their use in some bioapplications are shown in Table 1. In addition there is a lot of research devoted only to BNPs preparation using silica encapsulation without real application in bioanalysis [85–87]. It is important to mention the possible variations in size and charge, the possibility to dry and dissolve without a shift of properties, high stability against changing of external conditions and possibility to precisely vary the composition of active groups on the surface.

Application of QD-multiloaded labels is only at the beginning and is not restricted only to the area of synthesis of labels. It is clear that challenges will arise from the conjugation side. Also the enhancement of the label size will need more attention from surface chemistry to prevent nonspecific interactions. Surmounting these obstacles, multiloaded labels have bright possibilities. This

TABLE 1 Types of Silica-Based Bifunctional Nanocomposites-Containing Magnetic Nanoparticles (Orange Dot (Light Grey in Print Versions)) and Quantum Dots (Red Dot (Dark Grey in Print Versions))

Type of Bifunctional Nanocomposites	Application in Bioanalysis
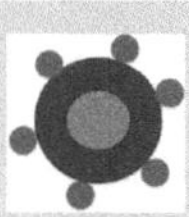	Lung cancer cells detection in pleural effusion
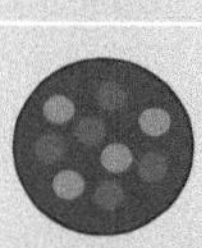	Immunofluorescence assay on microplate for detection of antibodies in solution by binding with secondary antibodies labelled with BNPs Cell-membrane labelling
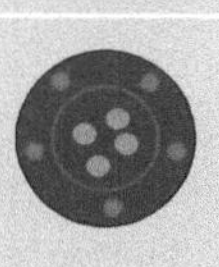	The investigation of cellular uptake of BNPS covered with thermosensitive polymer for further cytosolic drug-delivery research

gives perspectives for application of QD-loaded labels not only in few examples of assays (current-day situation), but also in a dramatically wider perspective, with validation on real samples, to be produced and commercially available in the future.

Despite historic achievements in the field of label-free bioassays, labelling techniques will continue to play a leading role in this field. Nanostuctures multiloaded with QDs offer very elegant ways of interfacing biomolecular recognition events with inherent signal amplification. The characteristics such as variability of QDs' amount, their distribution inside the nanostructures, density and composition of functional groups on the surface (with the exception of dendrimer-based structures), size and stability, availability of QDs for external substances (for example energy acceptors), possibility of controlling and tailoring their properties in a very predictable and precise manner meet the needs of analytical applications. Next to high variability of QDs' signal-causing characteristics (composition, size, and so on), this provides wide possibilities to optimize the labels' properties and allows to obtain high sensitivity. Because association in structure does not appreciably change the properties of the QDs, the same readers as for single QDs can be applied.

ACKNOWLEDGEMENT

The work was supported by Russian Science foundation (project 14-13-00229).

REFERENCES

[1] J. Wang, Electrochemical biosensors: towards point-of-care cancer diagnostics, Biosens. Bioelectron. 21 (2006) 1887–1892.

[2] F. Ricci, G. Adornetto, G. Palleschi, A review of experimental aspects of electrochemical immunosensors, Electrochim. Acta 84 (2012) 74–83.

[3] Y. Wan, Y. Su, X. Zhu, G. Liu, C. Fan, Development of electrochemical immunosensors towards point of care diagnostics, Biosens. Bioelectron. 47 (2013) 1–11.

[4] C. Ciminelli, C.M. Campanella, F. Dell'Olio, C.E. Campanella, M.N. Armenise, Label-free optical resonant sensors for biochemical applications, Progr. Quant. Electron. 37 (2013) 51–107.

[5] Y.S. Skibina, V.V. Tuchin, V.I. Beloglazov, G. Steinmeyer, J. Bethge, R. Wedell, N. Langhoff, Photonic crystal fibres in biomedical investigations, Quant. Electron. 41 (2011) 284–301.

[6] P.S.J. Russell, Photonic crystal fibers, J. Lightwave Technol. 24 (2006) 4729–4749.

[7] J. Villatoro, J. Zubia, New perspectives in photonic crystal fibre sensors, Opt. Laser Technol. 78 (2016) 67–75.

[8] A.V. Malinin, A.A. Zanishevskaja, V.V. Tuchin, Y.S. Skibina, I.Y. Silokhin, Photonic crystal fibers for food quality analysis, Proc. SPIE 8427 (2012) 842746–842748.

[9] P.S.J. Russell, Photonic crystal fibers, Science 299 (2003) 358–362.

[10] L. Rindorf, O. Bang, Sensitivity of photonic crystal fiber grating sensors: biosensing, refractive index, strain, and temperature sensing, J. Opt. Soc. Am. B 25 (2008) 310–324.

[11] A.A. Zanishevskaya, A.V. Malinin, V.V. Tuchin, YuS. Skibina, I.Yu Silokhin, Photonic crystal waveguide biosensor, J. Innov. Opt. Health Sci. 6 (2013), 1350008 (6 pages).

[12] A.V. Malinin, YuS. Skibina, V.V. Tuchin, M.V. Chainikov, V.I. Beloglazov, I.Yu Silokhin, A.A. Zanishevskaya, V.A. Dubrovskii, A.A. Dolmashkin, The use of hollow-core photonic crystal fibres as biological sensors, Quant. Electron. 41 (2011) 302–307.

[13] S.O. Konorov, A.M. Zheltikov, M. Scalora, Photonic-crystal fiber as a multifunctional optical sensor and sample collector, Opt. Express 13 (2005) 3454–3459.

[14] N. Skivesen, A. Têtu, M. Kristensen, J. Kjems, L.H. Frandsen, P.I. Borel, Photonic-crystal waveguide biosensor, Opt. Express 15 (2007) 3169–3176.

[15] J.M. Fini, Microstructure fibres for optical sensing in gases and liquids, Meas. Sci. Technol. 15 (2004) 1120–1128.

[16] A.M. Cubillas, S. Unterkofler, T.G. Euser, B.J.M. Etzold, A.C. Jones, P.J. Sadler, P. Wasserscheid, P.S.J. Russell, Photonic crystal fibres for chemical sensing and photochemistry, Chem. Soc. Rev. 42 (2013) 8629–8648.

[17] D. Yong, E. Lee, X. Yu, C.C. Chan, Fluorospectroscopy of dye-loaded liposomes in photonic crystal fibers, IEEE J. Sel. Top. Quantum Electron. 22 (2016) 4901006.

[18] L. Rindorf, J.B. Jensen, M. Dufva, L.H. Pedersen, P.E. Hoiby, O. Bang, Photonic crystal fiber long-period gratings for biochemical sensing, Opt. Express 14 (2006b) 8224–8231.

[19] A. Candiani, A. Bertucci, S. Giannetti, M. Konstantaki, A. Manicardi, S. Pissadakis, A. Cucinotta, R. Corradini, S. Selleri, Label-free DNA biosensor based on a peptide nucleic acid-functionalized microstructured optical fiber-Bragg grating, J. Biomed. Opt. 18 (2013) 057004.

[20] Z. He, F. Tian, Y. Zhu, N. Lavlinskaia, H. Du, Long-period gratings in photonic crystal fiber as an optofluidic label-free biosensor, Biosens. Bioelectron. 26 (2011) 774–4778.

[21] J.B. Jensen, P.E. Hoiby, G. Emiliyanov, O. Bang, L.H. Pedersen, A. Bjarklev, Selective detection of antibodies in microstructured polymer optical fibers, Opt. Express 13 (2005) 5883–5889.

[22] G. Emiliyanov, P.E. Hoiby, L.H. Pedersen, O. Bang, Selective serial multi-antibody biosensing with TOPAS microstructured polymer optical fibers, Sensors 13 (2013) 3242–3251.

[23] J.B. Jensen, L.H. Pedersen, P.E. Hoiby, L.B. Nielsen, T.P. Hansen, J.R. Folkenberg, J. Riishede, D. Noordegraaf, K. Nielsen, A. Carlsen, A. Bjarklev, Photonic crystal fiber based evanescent-wave sensor for detection of biomolecules in aqueous solutions, Opt. Lett. 29 (2004) 1974–1976.

[24] C.M.B. Cordeiro, M.A.R. Franco, G. Chesini, E.C.S. Barretto, R. Lwin, C.H. Brito Cruz, M.C.J. Large, Microstructured-core optical fiber for evanescent sensing applications, Opt. Express 14 (2006) 13056–13066.

[25] Y. Ruan, E.P. Schartner, H. Ebendorff-Heidepriem, P. Hoffmann, T.M. Monro, Detection of quantum-dot labelled proteins using soft glass microstructured optical fibers, Opt. Express 15 (2007) 17819–17826.

[26] P. Pinkhasova, H. Chen, J. Kanka, P. Mergo, H. Du, Nanotag-enabled photonic crystal fiber as quantitative surface-enhanced Raman scattering optofluidic platform, Appl. Phys. Lett. 106 (2015) 071106.

[27] N. Zohar, L. Chuntonov, G. Haran, The simplest plasmonic molecules: metal nanoparticle dimers and trimers, J. Photochem. Photobiol. C: Photochem. Rev. 21 (2014) 26–39.

[28] A. Khetani, V.S. Tiwari, A. Harb, H. Anis, Monitoring of heparin concentration in serum by Raman spectroscopy within hollow core photonic crystal fiber, Opt. Express 19 (2011) 15244–15254.

[29] I.Y. Goryacheva, E.S. Speranskaya, V.V. Goftman, D. Tang, S. De Saeger, Synthesis and bioanalytical applications of nanostructures multiloaded with quantum dots, Trends Anal. Chem. 66 (2015) 53–62.

[30] G.A. Posthuma-Trumpie, J.H. Wichers, M. Koets, L.B.J.M. Berendsen, A. van Amerongen, Amorphous carbon nanoparticles: a versatile label for rapid diagnostic (immuno)assays, Anal. Bioanal. Chem. 402 (2012) 593–600.

[31] K. Tanaka, S. Iijima (Eds.), Carbon Nanotubes and Graphene, second ed., Elsevier Ltd, 2014, 458 p.

[32] A. Jorio, G. Dresselhaus, M.S. Dresselhaus (Eds.), Carbon Nanotubes, Advanced Topics in the Synthesis, Structure, Properties and Applications, Topics in Applied Physics, vol. 111, 2008, 671 p.

[33] D. Cui, B. Pan, H. Zhang, F. Gao, R. Wu, J. Wang, R. He, T. Asahi, Self-assembly of quantum dots and carbon nanotubes for ultrasensitive DNA and antigen detection, Anal. Chem. 80 (2008) 7996–8001.

[34] B. Chen, H. Zhang, C. Zhai, N. Du, C. Sun, J. Xue, D. Yang, H. Hueng, D. Zhang, Q. Xie, Y. Wu, Carbon nanotube-based magnetic-fluorescent nanohybrids as highly efficient contrast agents for multimodal cellular imaging, J. Mater. Chem. 20 (2010) 9895–9902.

[35] C. Hussain, S. Mitra, Micropreconcentration units based on carbon nanotubes (CNT), Anal. Bioanal. Chem. 399 (2011) 75–89.

[36] P. Yanez-Sedeno, J.M. Pingarron, J. Riu, F.X. Rius, Electrochemical sensing based on the carbon nanotubes, Trend Anal. Chem. 29 (2010) 939–953.

[37] K. Scida, P.W. Stege, G. Haby, G.A. Messina, C.D. García, Recent applications of carbon-based nanomaterials in analytical chemistry: critical review, Anal. Chim. Acta 691 (2011) 6–17.

[38] Y.Y. Zhang, W. Qin, H. Tang, F. Yan, L. Tan, Q. Xie, M. Ma, Y. Zhang, S.Z. Yao, Efficient assembly of multi-walled carbon nanotube-CdSe/ZnS quantum dot hybrids with high biocompatibility and fluorescence property, Coll. Surf. B: Biointerfaces 87 (2011) 346–352.

[39] S. Banerjee, T. Hemraj-Benny, S.S. Wong, Covalent surface chemistry of single-walled carbon nanotubes, Adv. Mater. 17 (2005) 17–29.

[40] J.M. Haremza, M.A. Hahn, T.D. Krauss, Attachment of single CdSe nanocrystals to individual single-walled carbon nanotubes, Nano Lett. 2 (2002) 1253–1258.

[41] J.T. Sun, C.Y. Hong, C.Y. Pan, Surface modification of carbon nanotubes with dendrimers or hyperbranced polymers, Polym. Chem. 2 (2011) 998–1007.

[42] S.N. Ding, J.J. Xu, H.Y. Chen, Enhanced solid-state electrochemiluminescence of CdS nanocrystals composited with carbon nanotubes in H_2O_2 solution, Chem. Commun. 34 (2006) 3631–3633.

[43] G. Jie, J. Zhang, D. Wang, C. Cheng, H.Y. Chen, J.J. Zhu, Electrochemiluminescence immunosensor based on CdSe nanocomposites, Anal. Chem. 80 (2008) 4033–4039.

[44] N. Gan, J. Zhou, P. Xiong, F. Hu, Y. Cao, T. Li, Q. Jiang, An ultrasensitive electrochemiluminescent immunoassay for aflatoxin M1 in milk, based on extraction by magnetic graphene and detection by antibody-labeled CdTe quantum dots-carbon nanotubes nanocomposites, Toxins 5 (2013) 865–883.

[45] S. Chaudhary, J.H. Kim, K.V. Singh, M. Ozkan, Fluorescence microscopy visualization of single-walled carbon nanotubes using semiconductor nanocrystals, Nano Lett. 4 (2004) 2415–2419.

[46] G.B. Kim, B. Ramaraj, K.R. Yoon, A new strategy to assemble CdSe/ZnS quantum dots with multi-walled carbon nanotubes for potential application in imaging and photosensitization, Appl. Surf. Sci. 258 (2011) 1027–1032.

[47] N. Jia, Q. Lian, Z. Tian, X. Duan, M. Yin, L. Jing, S. Chen, H. Shen, M. Gao, Decorating multi-walled carbon nanotubes with quantum dots for construction of multi-color fluorescent nanoprobes, Nanotechnology 21 (2010) 045606.

[48] F. Liu, W. Deng, Y. Zhang, S. Ge, J. Yu, X. Song, Application of ZnO quantum dots dotted carbon nanotube for sensitive electrochemiluminescence immunoassay based on simply electrochemical reduced Pt/Au alloy and a disposable device, Anal. Chim. Acta 818 (2014) 46–53.

[49] A.M. Caminade, J.P. Majoral, Dendrimers and nanotubes: a fruitful association, Chem. Soc. Rev. 39 (2010) 2034–2047.

[50] J. Tian, H. Zhai, M. Liu, Y. Chen, X. Quan, Detection of influenza A virus based on fluorescence resonance energy transfer from quantum dots to carbon nanotubes, Anal. Chim. Acta 723 (2012) 83–87.

[51] G. Decher, Fuzzy nanoassemblies: toward layered polymeric multicomposites, Science 277 (1997) 1232–1237.

[52] Y. Chen, B. Jiang, Y. Xiang, Y. Chai, R. Yuan, Aptamer-based highly sensitive electrochemiluminescent detection of thrombin via nanoparticle layer-by-layer assembled amplification labels, Chem. Commun. 47 (2011) 7758–7760.

[53] G.B. Sukhorukov, E. Donath, H. Lichtenfeld, H. Knippel, M. Knippel, A. Budde, H. Mohwald, Layer-by-layer self assembly of polyelectrolytes on colloidal particles, Colloids Surf. A 137 (1998) 253–266.

[54] M. Adamczak, H.J. Hoel, G. Gaudernack, J. Barnasz, K. Szczepanowich, P. Warszynski, Polyelectrolyte multilayer capsules with quantum dots for biomedical application, Colloid Surf. 90 (2012) 211–216.

[55] A.N. Generalova, V.A. Oleinikov, M.M. Zarifullina, E.V. Lankina, S.V. Sizova, M.V. Artemyev, V.P. Zubov, Optical sensing quantum dot-labeled polyacrolein particles prepared by layer-by-layer deposition technique, J. Coll. Int. Sci. 357 (2011) 265–272.

[56] Y. Xiang, H. Zhang, B. Jiang, Y. Chai, R. Yuan, Quantum dot layer-by-layer assemblies as signal amplification labels for ultrasensitive electronic detection of uropathogens, Anal. Chem. 83 (2011) 4302–4306.

[57] B. Zhang, D. Tang, I.Y. Goryacheva, R. Niessner, D. Knopp, Anodic-stripping voltammetric immunoassay for ultrasensitive detection of low-abundance proteins using quantum dot aggregated hollow microspheres, Chem. Eur. J. 19 (2013) 2496–2503.

[58] W.Y. Cai, L.D. Feng, S.H. Liu, J.J. Zhu, Hemoglobin-CdTe-$CaCO_3$@polyelectrolytes 3D architecture: fabrication, characterization, and application in biosensing, Adv. Funct. Mater. 18 (2008) 3127–3136.

[59] S.M. Grayson, J.M.J. Frechet, Convergent dendrons and dendrimers: from synthesis to applications, Chem. Rev. 101 (2001) 3819–3867.

[60] Y. Higuchi, C. Wu, K.L. Chang, K. Irie, S. Kawakami, F. Yamashita, M. Hashida, Polyamidoamine dendrimer-conjugated quantum dots for efficient labeling of primary cultured mesenchymal stem cells, Biomaterials 32 (2011) 6676–6682.

[61] D.A. Geraldo, E.F. Duran-Lara, D. Aguayo, R.E. Cachau Jaime Tapia, R. Esparza, M.J. Yacaman, F.D. Gonzalez-Nilo, L.S. Santos, Supramolecular complexes of quantum dots and a polyamidoamine (PAMAM)-folate derivative for molecular imaging of cancer cells, Anal. Bioanal. Chem. 400 (2011) 483–492.

[62] F. Divsar, H. Ju, Electrochemiluminescence detection of near single DNA molecules by using quantum dots–dendrimer nanocomposites for signal amplification, Chem. Commun. 47 (2011) 9879–9881.

[63] G. Jie, L. Wang, J. Yuan, S. Zhang, Versatile electrochemiluminescence assays for cancer cells based on dendrimer/CdSe/ZnS quantum dot nanoclusters, Anal. Chem. 83 (2011) 3873–3880.

[64] M. Akin, R. Bongartz, J.G. Walter, D.O. Demirkol, F. Stahl, S. Timur, T. Scheper, PAMAM-functionalized water soluble quantum dots for cancer cell targeting, J. Mater. Chem. 22 (2012) 11529–11536.

[65] Z. Cheng, D.L.J. Thorek, A. Tsourkas, Gadolinium-conjugated dendrimer nanoclusters as a tumor-targeted T1 magnetic resonance imaging contrast agent, Angew. Chem. Int. Ed. 49 (2010) 346–350.

[66] G. Jie, J. Yuan, J. Zhang, Quantum dots-based multifunctional dendritic superstructure for amplified electrochemiluminescence detection of ATP, Biosens. Bioelectron. 31 (2012) 69–76.

[67] Y. Zeng, C. Tang, H. Wang, J. Jiang, M. Tian, G. Shen, R. Yu, A novel density-tunable nanocomposites of CdTe quantum dots linked to dendrimer-tethered multi-wall carbon nanotubes, Spectrochim. Acta A 70 (2008) 966–972.

[68] D. Tang, L. Hou, R. Niessner, M. Xu, Z. Gao, D. Knopp, Multiplexed electrochemical immunoassay of biomarkers using metal sulfide quantum dot nanolabels and trifunctionalized magnetic beads, Biosens. Bioelectron. 46 (2013) 37–43.

[69] D. Yamamoto, T. Koshiyama, S. Watanabe, M.T. Miyahara, Synthesis and photoluminescence characterization of dendrimer-encapsulated CdS quantum dots, Colloids Surf. A: Physicochem. Eng. Aspects 411 (2012) 12–17.

[70] M. Algarra, B.B. Campos, M.S. Miranda, J.C.G. Esteves da Silva, CdSe quantum dots capped PAMAM dendrimer nanocomposites for sensing nitroaromatic compounds, Talanta 83 (2011) 1335–1340.

[71] C.L. Feng, X.H. Zhong, M. Steinhart, A.M. Caminade, J.P. Majoral, W. Knoll, Functional quantum dot/dendrimer nanotubes for sensitive detection of DNA hybridization, Small 4 (2008) 566–571.

[72] T. Nann, P. Mulvaney, Single quantum dots in spherical silica particles, Angew. Chem. Int. Ed. 43 (2004) 5393–5396.

[73] P. Yang, N. Murase, Size-tunable highly luminescent SiO_2 particles impregnated with number-adjusted CdTe nanocrystals, Chem. Phys. Chem. 11 (2010) 815–821.

[74] P. Yang, L. Zhang, X. Li, Y. Zhang, N. Liu, R. Zhang, Surface modification of SiO_2 beads with multiple hydrophobic quantum dots for bioapplications, J. Non-Cryst. Solids 358 (2012) 3069–3073.

[75] C. Kirchner, T. Liedl, S. Kudera, T. Pellegrino, A.M. Javier, H.E. Gaub, S. Stolzle, N. Fertig, W.J. Parak, Cytotoxicity of colloidal CdSe and CdSe/ZnS nanoparticles, Nano Lett. 5 (2005) 331–338.

[76] Y. Zhu, Z. Li, M. Chen, H.M. Cooper, Q.G.M. Lu, Z.P. Xu, Synthesis of robust sandwich-like SiO_2@CdTe@SiO_2 fluorescent nanoparticles for cellular imaging, Chem. Mater. 24 (2012) 421–423.

[77] W. Stöber, A. Fink, E. Bohn, Controlled growth of monodisperse silica spheres in the micron size range, J. Colloid Interface Sci. 26 (1968) 62–69.

[78] P. Yang, N. Murase, J. Yu, SiO_2 beads with quantum dots: preparation and stability investigation for bioapplications, Colloids Surf. A: Physicochem. Eng. Aspects 385 (2011) 159–165.

[79] P. Yang, N. Murase, M. Suzuki, C. Hosokawa, K. Kawasaki, T. Kato, T. Taguchi, Bright, non-blinking, and less-cytotoxic SiO_2 beads with multiple CdSe/ZnS nanocrystals, Chem. Commun. 46 (2010) 4595–4597.

[80] Y. Piao, A. Burns, J. Kim, U. Wiesner, T. Hyeon, Designed fabrication of silica-based nanostructured particle systems for nanomedicine applications, Adv. Funct. Mater. 18 (2008) 3745–3758.

[81] S. Wang, C. Li, P. Yang, M. Ando, N. Murase, Silica encapsulation of highly luminescent hydrophobic quantum dots by two-step microemulsion method, Colloids Surf. A: Physicochem. Eng. Aspects 395 (2012) 24–31.

[82] Y. Kobayashi, H. Matsudo, T. Nakagawa, Y. Kubota, K. Gonda, N. Ohuchi, In-vivo fluorescence imaging technique using colloid solution of multiple quantum dots/silica/poly(ethylene glycol) nanoparticles, J. Sol-Gel Sci. Technol. 66 (2013) 31–37.
[83] Y. Yang, L. Jing, X. Yu, D. Yan, M. Gao, Coating aqueous quantum dots with silica via reverse microemulsion method: toward size-controllable and robust fluorescent nanoparticles, Chem. Mater. 19 (2007) 4123–4128.
[84] M. Darbandi, R. Thomann, T. Nann, Single quantum dots in silica spheres by microemulsion synthesis, Chem. Mater. 17 (2005) 5720–5725.
[85] K. Osseo-Asare, F.J. Arriagadaf, Preparation of SiO_2 nanoparticles in a non-ionic reverse micellar system, Colloids Surf. 50 (1990) 321–339.
[86] S.T. Selvan, P.K. Patra, C.Y. Ang, J.Y. Ying, Synthesis of silica-coated semiconductor and magnetic quantum dots and their use in the imaging of live cells, Angew. Chem. Int. Ed. 46 (2007) 2448–2452.
[87] J. Guo, W. Yang, C. Wang, J. He, J. Chen, Poly(N-isopropylacrylamide)-coated luminescent/magnetic silica microspheres: preparation, characterization, and biomedical applications, Chem. Mater. 18 (2006) 5554–5562.
[88] E.E. Lees, T.L. Nguyen, A.H.A. Clayton, P. Mulvaney, The preparation of colloidally stable, water-soluble, biocompatible, semiconductor nanocrystals with a small hydrodynamic diameter, ACS Nano 3 (2009) 1121–1128.
[89] I.Y. Goryacheva, Formats of rapid immunotests – current-day formats, perspectives, pros and cons, in: I.Y. Goryacheva (Ed.), Rapid Immunotests for Clinical, Food and Environmental Applications vol. 72, 2016, pp. 33–78.
[90] I.Y. Goryacheva, Rapid tests progress through the years, in: I.Y. Goryacheva (Ed.), Rapid Immunotests for Clinical, Food and Environmental Applications vol. 72, 2016, pp. 5–32.

Chapter 7

Conclusion and Outlook

Chapter Outline

Development of immunoassays and rapid tests began with natural components, and decades of progress resulted in a variety of effective tools meeting modern requirements. The following specifications can be considered 'ideal' for rapid tests:

- Sample-based testing, ie, making the test applicable to process samples without their preparation. This is especially true for liquid samples, such as serum, urine, saliva, environmental water, milk, juice, etc.
- Exclusion of washing, separation and other handling steps, for example, microfluidic or lateral flow immunoassays. It provides not only an opportunity to use untrained personnel and the lack of uncertainty as a result of handling steps, but also simplify the documentation of the results.
- Fast test processing (rapid binding kinetics and quick measurement), which is important for clinical assays and for perishable goods as well as special tasks such as roadside screening for drugs of abuse in saliva for traffic safety. To shorten the time of analysis, capturing antibody can be attached to a developed surface (micro-heterogeneous systems, nanostructures) to ensure rapid transfer of molecules or conjugates to be captured.
- Small sample volumes, which is more important for clinical assays, especially in pediatrics.
- Sensitivity (relevant detection limits) and specificity (the use of high-specific antibodies) of analysis.
- Reproducibility (ratiometric or related measurement for correction of medium interferences).
- Robustness (stability of outcomes is independent from individual user or environment).
- Flexibility (universal format for many analytes).
- Multiplexity (simultaneous measurement of several analytes).
- Stability (long-term storage).
- Absence of toxic solvents or substances to avoid problems with utilization in distant locations.
- Possibility to document assay outcomes and assay conditions.

Comprehensive Analytical Chemistry, Vol. 72. http://dx.doi.org/10.1016/bs.coac.2016.03.011

Components and principles of modern immunotests use a variety of combinatorial strategies to ensure the best performance:

- Lateral flow immunoassay is a combination of chromatography (separation of components of a sample based on differences in their movement through a sorbent) and immunochemical reactions.
- Electrodes and photon crystal fibres combine a solid support for immunoreagents binding and signal processing.
- Lab on chip is a combination of microfluidic system (in ideal case – sample clean-up, dilution, mixing with reagents and incubation (if necessary)) and immunochemical method.

It is interesting to note that together with simplification (easy-to-use, out-of-lab decentralization) strategy of rapid tests progress, the opposite direction develops as well. This involves the use of rapid tests in complex test systems in laboratories. For example, rapid tests can serve as components of highly specified platforms with advanced-facilitative technologies in the form of readers, specifically designed sample processing devices and cartridge approaches with onboard functionality – essentially creating laboratory analysers with the rapid immunotests, such as lateral flow strips, at their core. Besides, applications in the developed world have taken advantage of the improved performance to gradually transform rapid tests into a true laboratory-based system.

Transition from qualitative to quantitative assay resulted in improvement of sensitivity. For example, the limit of detection for lateral flow immunoassay has increased by 3–50 times. These gains were achieved by more sensitive detection of weak staining via alternative methods of label detection, by measuring the magnitude of signal decreases rather than signal disappearance in the test zones of competitive systems [1,2].

Common advantages of immunochemical methods include the possibility to adapt once optimized set of immunoreagents for different tests and methods or to develop a new format, principle or setting for detection of one analyte and then apply it to other substances. Usually, new formats of rapid tests are first developed for the clinical area (point-of-care testing) and then applied to the food and feed control areas and finally to the field of environmental control. The immunoassay format changes in parallel: from mostly sandwich format for high-molecular-weight analytes of clinical interest to a competitive format for mostly low-weight analytes in ecological monitoring [3].

Most publications about rapid tests and commercial rapid tests are related to the clinical area. The development of rapid tests for food safety and quality control is a relatively new field of research. There have been several reviews of recent progress in the area of food analysis [1,4,5]. Most of them described research related to the development of systems capable of detecting a group of food contaminants or a particular compound, and discussed factors related to increasing the speed and sensitivity of the proposed assays. Significantly, fewer amount of publications concerning rapid tests for ecotoxicological items could be found [6–8].

A very promising direction, particularly for health-related assays, is application of cell phones as component of rapid tests. If equipped with a reader platform, cell phones can solve a series of tasks, including reading the signal, processing and quantifying it, sending the outcomes to centralized laboratory or physician's office for expert interpretation, documenting and keeping data or signalling the alarm. Application of smartphones as signal readers and processors makes it easier to transfer assays out of laboratories. Current status of wireless telecommunication technologies exhibits a promising potential to be utilized for different applications even in the least developed parts of the world. With more than 5 billion subscriptions worldwide, the cell phone, as an ubiquitous platform, can be utilized for imaging, sensing, processing and communicating rapid tests outcomes; first of all health-related data in field settings using already embedded digital components [9,10]. But the first part of assay – signal production as a result of immunoassay itself – is still an operator's liability. So the main efforts should be focused on stability, reproducibility and simplicity of the immunochemical parts of assay, including sampling, sample preparation and assay realization. This demands increasing the accuracy of chemical steps. For some points, inclusion of calibrators or positive and negative controls for the application of ratiometric or associated signal evaluation methods is important.

Considering the breadth of the field that the rapid tests may offer, their coverage in this book is by no means complete.

ACKNOWLEDGEMENT

The work was supported by Russian Science Foundation (project 14-13-00229).

REFERENCES

[1] B.B. Dzantiev, N.A. Byzova, A.E. Urusov, A.V. Zherdev, Immunochromatographic methods in food analysis, Trends Anal. Chem. 55 (2014) 81–93.

[2] D. Pyo, J. Yoo, New trends in fluorescence immunochromatography, J. Immunoassay Immunochem. 33 (2012) 203–222.

[3] I.Y. Goryacheva, P. Lenain, S. De Saeger, Nanosized labels for rapid immunotests: new trends and developments, Trends Anal. Chem. 46 (2013) 30–46.

[4] C.M. Maragos, M. Busman, Rapid and advanced tools for mycotoxin analysis: a review, Food Addit. Contam. Part A Chem. 27 (2010) 688–700.

[5] I.Y. Goryacheva, S. de Saeger, S.A. Eremin, C. van Peteghem, Immunochemical methods for rapid mycotoxin detection: evolution from single to multiple analyte screening: a review, Food Addit. Contam. A 24 (2007) 1169–1183.

[6] Y.R. Guo, S.Y. Liu, W.J. Gui, G.N. Zhu, Gold immunochromatographic assay for simultaneous detection of carbofuran and triazophos in water samples, Anal. Biochem. 389 (2009) 32–39.

[7] S. Girotti, S. Eremin, A. Montoya, M.J. Moreno, P. Caputo, M.D. Elia, L. Ripani, F.S. Romolo, E. Maiolini, Development of a chemiluminescent ELISA and a colloidal gold-based LFIA for TNT detection, Anal. Bioanal. Chem. 396 (2010) 687–695.

[8] Y.M. Kim, S.W. Oh, S.Y. Jeong, D.J. Pyo, E.Y. Choi, Development of an ultrarapid one-step fluorescence immunochromatographic assay system for the quantification of microcystins, Environ. Sci. Technol. 37 (2003) 1899–1904.

[9] O. Mudanyali, S. Dimitrov, U. Sikora, S. Padmanabhan, I. Navruz, A. Ozcan, Integrated rapid-diagnostic-test reader platform on a cellphone, Lab Chip 12 (2012) 2678–2686.

[10] D.S. Lee, B.G. Jeon, C. Ihm, J.K. Park, M.Y. Jung, A simple and smart telemedicine device for developing regions: a pocket-sized colorimetric reader, Lab Chip 11 (2011) 120–126.

Index

'*Note*: Page numbers followed by "f" indicate figures and "t" indicate tables.'

C

G

H

M

N

O

S

Kolibri 360 GmbH
Ferdinand-Jühlke-Straße 7
99095 Erfurt, Deutschland
produktsicherheit@kolibri360.de